3000 800066 54432
St. Louis Community College

D0857630
314-513-4514

Palgrave Macmillan Asian Business Series

Series editor: **Harukiyo Hasegawa** is Professor at Doshisha Business School, Kyoto, Japan, and Honourable Research Fellow at the University of Sheffield's School of East Asian Studies, where he was formerly Director of the Centre for Japanese Studies.

The Palgrave Macmillan Asian Business Series seeks to publish theoretical and empirical studies that contribute forward-looking social perspectives on the study of management issues not just in Asia, but by implication elsewhere. The series specifically aims at the development of new frontiers in the scope, themes and methods of business and management studies in Asia, a region which is seen as key to studies of modern management, organization, strategies, human resources and technologies.

The series invites practitioners, policymakers and academic researchers to join us at the cutting edge of constructive perspectives on Asian management, seeking to contribute towards the development of civil societies in Asia and further afield.

Titles include:

Sten Söderman (*ed.*)
EMERGING MULTIPLICITY
Integration and Responsiveness in Asian Business Development

Oliver H. M. Yau and Raymond P. M. Chow (*eds*)
HARMONY VERSUS CONFLICT IN ASIAN BUSINESS
Managing in a Turbulent Era

Hiroaki Miyoshi and Yoshifumi Nakata (*eds*)
HAVE JAPANESE FIRMS CHANGED?
The Lost Decade

Glenn D. Hook and Harukiyo Hasegawa (*eds*)
JAPANESE RESPONSES TO GLOBALIZATION IN THE 21st CENTURY
Politics, Security, Economics and Business

Diane Rosemary Sharpe and Harukiyo Hasegawa (*eds*)
NEW HORIZONS IN ASIAN MANAGEMENT
Emerging Issues and Critical Perspectives

Hiroaki Miyoshi and Masanobu Kii (*eds*)
TECHNOLOGICAL INNOVATION AND PUBLIC POLICY
The Automotive Industry

Tan Yi
THE OIL AND GAS SERVICE INDUSTRY IN ASIA
A Comparison of Business Strategies

Palgrave Macmillan Asian Business Series
Series Standing Order ISBN 978–1–4039–9841–5

You can receive future titles in this series as they are published by placing a standing order. Please contact your bookseller or, in case of difficulty, write to us at the address below with your name and address, the title of the series and the ISBN quoted above.

Customer Services Department, Macmillan Distribution Ltd, Houndmills, Basingstoke, Hampshire RG21 6XS, England

Technological Innovation and Public Policy

The Automotive Industry

Edited by

Hiroaki Miyoshi

and

Masanobu Kii

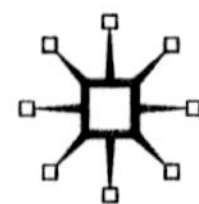

First published 2011 by
PALGRAVE MACMILLAN

Palgrave Macmillan in the UK is an imprint of Macmillan Publishers Limited, registered in England, company number 785998, of Houndmills, Basingstoke, Hampshire RG21 6XS.

Palgrave Macmillan in the US is a division of St Martin's Press LLC, 175 Fifth Avenue, New York, NY 10010.

Palgrave Macmillan is the global academic imprint of the above companies and has companies and representatives throughout the world.

Palgrave® and Macmillan® are registered trademarks in the United States, the United Kingdom, Europe and other countries.

ISBN: 978–0–230–23076–7 hardback

This book is printed on paper suitable for recycling and made from fully managed and sustained forest sources. Logging, pulping and manufacturing processes are expected to conform to the environmental regulations of the country of origin.

A catalogue record for this book is available from the British Library.

Library of Congress Cataloging-in-Publication Data

Technological innovation and public policy : the automotive industry / edited by Hiroaki Miyoshi and Masanobu Kii.
p. cm.
Includes index.
ISBN 978–0–230–23076–7 (hardback)
1. Automobile industry and trade – Government policy – Japan.
2. Automobile industry and trade – Technological innovations – Japan.
3. Transportation and state – Japan. I. Miyoshi, Hiroaki, 1960–
II. Kii, Masanobu.

HD9710.J32T43 2011
338.4'76292220952—dc22 2011012068

10 9 8 7 6 5 4 3 2 1
20 19 18 17 16 15 14 13 12 11

Printed and bound in the United States of America

Contents

Tables

Figures

Preface

This book presents the results of research conducted at Doshisha University's Institute for Technology, Enterprise and Competitiveness (ITEC) and supported by the 'Synthetic Research on Technology, Enterprise and Competitiveness', a twenty-first-century COE program sponsored by Japan's Ministry of Education, Culture, Sports, Science and Technology, as well as a Grant-in-Aid for Scientific Research on Priority Area (B) (No. 21310099) from the Japan Society for the Promotion of Science (JSPS).

A Japanese-language version of this book, titled *Technological Innovation in the Automotive Industry and Economic Welfare*, was published in March 2008 by Hakuto-Shobo Publishing Company. In preparing this English translation, we have updated a significant portion of the content to reflect the prevailing conditions, and we have added discussions of our most recent research, including studies of safety-related intelligent transport systems.

The contributors to this book are the members of a single organizational research unit within ITEC. The research objectives of ITEC are to promote innovation and societal well-being, and its operational style – including the integration of natural and social sciences, frequent international collaborations, and tight-knit cooperation between industry, academia, and government – has made it one of the leading institutions for research in the social sciences in Japan. The contributors to this book wholeheartedly support ITEC's research goals and its operating philosophy, and ITEC in turn depends on the hard work and creativity of the researchers whose work comprises this book.

There are a number of people we wish to thank for their invaluable support, both of this book and of the development of ITEC. In particular, we are grateful to Professor Harukiyo Hasegawa of Doshisha University, the Series Editor at Palgrave Macmillan, for giving us the opportunity to publish the results of our research. Professor Yoshifumi Nakata, Director-General of ITEC, assisted with many tasks in the organizing and publishing of this book. Professor Tadashi Yagi of Doshisha University provided valuable comments and advice on this book from the viewpoint of public economics. In conducting the research described in this book, we were fortunate to receive tutelage from representatives of automobile manufacturers regarding the latest technological trends and other up-to-date news from the automobile industry; in this regard we are particularly grateful to Eishi Ohno of Toyota Motor Corporation and to Toshio Yokoyama and Satoshi Hada at Honda R&D Co. Ltd. We are grateful to our publisher, Palgrave Macmillan, and in particular to our editor, Virginia Thorp, and our editorial assistant, Paul Milner, for keeping us on track toward the publication of this book. In addition, we thank Michiyo Tanaka at ITEC for her indefatigable assistance with the minutiae of the editing process.

We conclude by expressing our sincere hope that this volume will be a useful contribution to today's most urgent policy debates, including how best to respond to problems, such as global warming and traffic accidents, and how best to ensure a healthy and sustainable motorization process in the rapidly developing BRIC nations – Brazil, Russia, India, and China.

Hiroaki Miyoshi
Director of the Institute of Technology, Enterprise and
Competitiveness at Doshisha University
March 2011

Contributors

Yuko Akune is Associate Professor of Economics in the Faculty of Economics and Business Administration at Reitaku University in Japan. She worked at the GENDAI Advanced Studies Research Organization, a consulting firm founded by Toyota Motor Co. Ltd (2003–09), and at the University of Tsukuba (2009–10), before joining Reitaku University in 2010. Professor Akune conducted forecasts of new vehicle demand for three years while at the GENDAI Advanced Studies Research Organization. Her research interests include analyzing the impact of vehicle-related taxation systems and regulations on new vehicle demand and CO_2 emission. Recent publications co-authored by Professor Akune include 'Location Choice and Agglomeration Effect on Japanese Multinational Food Companies in East Asia and NAFTA/EU: Panel Data Analysis', *Journal of Applied Regional Science* (2007, in Japanese) and 'An Empirical Analysis of the Industrial Agglomeration of Food Industry from 1980 to 2000: Using Agglomeration and Co-Agglomeration Indexes by Ellison and Glaeser (1997)', *Studies in Regional Science* (2005, in Japanese).

Masanobu Kii is Associate Professor of Environmental Policies and Planning in the Faculty of Engineering of Kagawa University in Japan. He worked at the Institute for Transport Policy Studies (2000–03), the Japan Automobile Research Institute (2004–08), and the Research Institute of Innovative Technology for the Earth (2008–09) before joining Kagawa University in 2009. Professor Kii has conducted impact analyses of several transport-sector policies, including fuel economy regulations, public transportation policies, and urban land-use planning. His research interests include sustainability of energy and environment systems. Recent publications authored or co-authored by Professor Kii include 'An Integrated Evaluation Method of Accessibility, Quality of Life, and Social Interaction', *Environment and Planning B* (2007), 'Impact Assessment of Fuel-Efficient Technologies for Passenger Vehicles', *Energy and Resources* (2007, in Japanese), and 'Multiagent Land-Use and Transport Model for the Policy Evaluation of a Compact City', *Environment and Planning B* (2005).

Hiroaki Miyoshi is Director of the Institute for Technology, Enterprise and Competitiveness (ITEC) at Doshisha University in Japan and

Professor at the Graduate School of Policy and Management at Doshisha University. He worked at a major private-sector think tank for 10 years, conducting research and making policy recommendations on a broad range of issues as a senior researcher, before joining Doshisha University in 2003. His research interests include public economics and transport economics. His recent publications include *Have Japanese Firms Changed? The Lost Decade* (2011, co-edited with Y. Nakata), *Knowledge Asset Management and Organizational Performance* (2009, co-edited with Y. Shozugawa, in Japanese), *Technological Innovation in the Automotive Industry and Economic Welfare* (2008, co-edited with M. Tanishita, in Japanese), and 'Factors Associated with Safety of Passenger Cars', *IATSS Research* (2008, co-author).

Masayuki Sano is a consultant and founder of Libertas Terra Co. Ltd, a consulting firm specializing in energy and environmental issues. He has served as director of a major private-sector think tank for the past five years and has supervised a variety of research projects sponsored primarily by governmental organizations. He currently serves as a consultant to the Japan Automobile Manufacturers Association Inc. (JAMA). From 2006 to 2009, he was a visiting fellow at ITEC at Doshisha University. His research interests include energy and environmental issues in the global road-transportation sector. His recent publications include 'Genealogy of Automotive Technology Innovation and Technology Policy in Japan', in Miyoshi, H. and Tanishita, M. (eds) *Technological Innovation in the Automotive Industry and Economic Welfare* (2008, in Japanese).

Lin Sun is Associate Professor at the Shanghai Academy of Social Sciences (SASS) in China and a Visiting Fellow at the Institute for Technology, Enterprise and Competitiveness (ITEC) at Doshisha University in Japan. He has worked at SASS since 2002, conducting analyses of energy, environment, and transport policies related to automobiles. Recent publications, authored or co-authored by Professor Sun, include 'Numerical Analysis of Environmental and Energy Policies related to Automobiles in China: Evaluation by Dynamic Computable General Equilibrium Model', *Studies in Regional Science* (2006, in Japanese), 'The impact of Consumption Tax Reform for Passenger Car in China: Simulation Analysis by CGE Model', *Forum of International Development Studies* (2007, in Japanese), and 'China's Automobile Traffic Issues and Auto Technology Policy', in Miyoshi, H. and Tanishita, M. (eds) *Technological Innovation in the Automotive Industry and Economic Welfare* (2008, in Japanese).

Masayoshi Tanishita is Professor in the Department of Civil and Environmental Engineering in the Faculty of Science and Engineering at Chuo University in Japan. Since completing his Doctor of Engineering at the University of Tokyo, he has worked at Chuo University for 14 years, conducting research on taxation and regulation policies regarding transport and land use. His research interests include transport and urban economics. His recent publications include *Technological Innovation in the Automotive Industry and Economic Welfare* (2008, co-edited with H. Miyoshi, in Japanese), 'Factors Associated with Safety of Passenger Cars', *IATSS Research* (2008, co-author), and 'Impact Analysis of Car-related Taxes on Fuel Consumption in Japan', *Journal of Transport Policy and Economics* (2003, co-author).

1
Introduction

Hiroaki Miyoshi and Masanobu Kii

1.1 Objectives of this book

The automobile is the most successful technology of the twentieth century. Its origins are known to lie in a steam-engine-powered vehicle, designed to transport cannons, invented by the French military engineer Cugnot in 1769. Later, in the second half of the nineteenth century, gasoline-powered vehicles with almost the same structure as those we know today were invented, and began to appear on the market in Germany, France, and the US at the end of the nineteenth century.

Until the beginning of the twentieth century, the automobile was a status symbol for the privileged classes. The popularization of automobiles, which led to the prosperity of today's automobile industry, was triggered by the Ford Model T, introduced by Henry Ford. The Model T was launched in 1908, and, thanks to the cost reductions realized by its innovative manufacturing system and its various mechanisms for simplifying driving, became the best-selling vehicle of all time. Inspired by the Ford Model T, the popularization of automobiles then took off in Europe, and motorization advanced globally in the 1920s. As of 2007, approximately 73 million new cars are produced every year, and approximately 950 million cars are in operation worldwide. The automobile is now an essential means of production and a critical enabler of our daily lives.

And yet the tremendous conveniences afforded by motorization have not come without cost. A host of problems related to the mass consumption of automobiles emerged in the latter half of the twentieth century and have yet to be conclusively resolved; among these we may list traffic accidents, air pollution, and traffic congestion. With regard to the traffic-accident situation in Japan, the number of fatalities related to traffic accidents peaked in the 1970s and has been on the decline ever since, thanks to safety measures imposed both on individual vehicles and on the roadway infrastructure. On the other hand, the frequency of non-fatal accidents and the number of injuries sustained still remain high, although both have decreased since 2005. As for air pollution, although the situation has been significantly improved by strengthened environmental regulations (including an expansion in the range of pollutants considered and the vehicles targeted) since the establishment of the Air Pollution Control Law in 1968, some areas of metropolitan regions still fail to meet environmental standards. Similarly, considering the problem of traffic congestion, it has been estimated that the time lost to traffic congestion in Japan amounts to some 3.8 billion hours per year nationwide, equivalent to a loss of approximately 12 trillion yen per year (as of March 2003, based on estimates by Japan's Ministry of Land, Infrastructure, Transport and Tourism). On top of these problems, we face the urgent challenge of addressing global warming, an issue which may threaten the continued existence of human civilization. The Framework Convention on Climate Change was established at the Earth Summit in 1992, while the Kyoto Protocol, which sets reduced greenhouse gas emissions targets for all developed countries, was adopted by the Third Conference of the Parties (COP3) in 1997. According to this protocol, Japan is required to reduce greenhouse gas emissions by 6 per cent (compared with 1990 levels) in the years between 2008 and 2012. CO_2 emissions from the transport sector account for approximately 20 per cent of all CO_2 emissions nationwide, with automobiles responsible for 88 per cent of transport sector emissions. According to 2008 figures for greenhouse gas emissions (Ministry of the Environment, Japan, 2010), emissions from passenger cars are on the rise (+31 per cent over the base year) while emissions from trucks are on the decline (–13 per cent). Among other factors, emission levels from private passenger cars remain high, increasing by 36 per cent over the base year, although these levels did recently take a downward turn.

In order to respond to these negative legacies of the twentieth century, car-makers have actively pursued the development of new technologies, and many of their recent achievements, including power system-related technologies such as hybrid engines and fuel cells, as well as travel-related technologies such as intelligent transportation systems, are now garnering significant attention.

In this book, after reviewing the history of Japanese automotive technology policies since the end of World War II, we consider how the new automotive technologies of the twenty-first century may best be put to use. The primary objective of this book is to investigate how innovative vehicle technology may be most effectively utilized, both to improve societal economic welfare and to address the problems of traffic accidents and of the global environment.

This book is the revised version of our earlier book, *Jidousya no Gijyutsu Kakushin to Keizai Kousei (Technological Innovation in the Automotive Industry and Economic Welfare)*, edited by H. Miyoshi and M. Tanishita and published in Japanese by Hakuto-Shobo Publishing Company in March 2008 (ISBN 978-4-561-96113-0 C3033).

1.2 Previous studies and the distinctive features of this book

Two representative volumes addressing problems related to automobile traffic in a comprehensive manner are Black and Nijkamp (2002) and Small and Verhoef (2007). The former of these works considers important global transportation issues, including environmental impact, business and economic considerations, societal reliance on automobiles, and new automotive technologies. Small and Verhoef (2007) provide an introduction to major research themes in contemporary transportation economics, such as travel demand, pricing, investment, and industrial organization of transportation providers. Greene *et al.* (1997) apply techniques from neoclassical economics to quantify the broad impact of automobiles on society. Sperling and Gordon (2010) examine the complex problems and challenges confronting the transportation sector from a multifaceted perspective, including consideration of vehicles, fuels, industries, consumer behavior, and government policy. The discussion is mainly based on the situation of the US, with some discussion of China.

In the area of environmental issues and fuel economy, Ryan and Turton (2007) consider long-term energy–economy–environment scenarios and identify the key technological developments required to mitigate the impact of passenger vehicles on climate change and to achieve energy security. This work also considers possible targets for policy support and examines some of the elements that contribute to the significant levels of uncertainty present in this field. Schafer *et al.* (2009) assess the factors affecting greenhouse gas emissions from passenger vehicle transportation, including the influence of personal and business choices, technologies and alternative fuels, and policies. These authors discuss the development of more sustainable transportation systems over the next 30 to 50 years, considering a broad spectrum of options.

With regard to regulation and taxation, a report from the National Research Council (2003) evaluates the need for the CAFE (Corporate

Average Fuel Economy) program, and the optimal structure and severity of the CAFE standards in the future, taking into account the implications of changes in motor vehicle technology, the distribution of vehicle sales, and a variety of other factors. Sevigny (1998) examines the role that an automobile emissions tax could play in reducing emissions in the US. This study contains detailed analyses of several aspects of such a proposal, including (1) the design of the tax; (2) behavioral responses that lead to emissions reductions, including reductions in vehicle miles traveled per household and the elimination of low-value, high-emissions vehicles; (3) the effect of the tax on the reduction of emissions; (4) the effect of the tax on households in different income quintiles; and (5) the emissions-reducing potential of a gasoline tax as compared with an emissions tax.

In the area of intelligent transportation systems (ITSs), Stough (2001) developed a new outcome-oriented methodology and applied it to a diverse set of ITS case studies. The cases studied include evaluation of electronic tolling, truck rollover warning systems, Advanced Traffic Information Systems (ATISs), variable message signs (VMSs), ITS-enhanced emergency management systems, and ITS bridge operations. Ghosh and Lee (2010) examined the successes and failures of ITSs in the past decade. They identified challenging problems that must be addressed in order to improve quality of life with consideration of environmental issues. Bekiaris and Nakanishi (2004) proposed new economic assessment techniques for ITSs that take into account the life cycles and cost structures of the systems, as well as a number of interrelated elements. This work also includes case studies covering a wide range of ITS technologies, including incident management, electronic toll collection, advanced driver assistance systems, and traveler information systems.

In comparison to these previous works, the uniqueness of this volume lies in its policy-oriented research and its theoretical analyses in the field of automotive technologies. First, automotive technological policies, and the process by which they are crafted, are reviewed in detail for the particular cases of Japan (Chapter 2) and China (Chapter 8). Readers will learn how automotive technology policy has been designed and implemented in order to mitigate or eliminate problems arising from the use of automobiles, including safety and environmental issues.

In addition to these surveys of concrete real-world policies, this volume also analyzes theoretical policies intended to promote the spread of technologies, to improve economic welfare, and to address a host of global problems; in the analyses reported in this book, the economic properties – such as *costs* and *externalities* – of new automobile technologies are analyzed within the framework of welfare economics. The book includes both theoretical contributions, which discuss analytical techniques, and empirical analyses, which apply abstract concepts and methods from welfare economics and microeconomics to the specific assessment of new automotive technologies.

Through this volume, we expect that readers will find not the detail, but the essence, of the way to discuss what policies are required to ensure that new automotive technologies are utilized effectively from the perspective of social welfare.

1.3 The structure of this book

This book consists of three main parts. The first part (Chapter 2) reviews the development of automotive technological policies in Japan. The second part (Chapters 3 to 7) analyzes the quantitative effects of automotive technological policies. Finally, the third part (Chapter 8) discusses automotive technological policies in China. In what follows we will briefly summarize each of these three parts.

In the latter half of the twentieth century, the Japanese automotive industry earned the world's respect and admiration for its efficient production systems – a fact well documented in any number of books, including *The Machine That Changed the World* (Womack *et al.*, 1990). However, few books have discussed the quality of Japanese cars from the perspective of government technology policies, a glaring deficiency in view of the critical role that such policies played in heightening the quality assurance standards of Japanese car-makers. Chapter 2 of this book is the first comprehensive historical survey of the interrelation between technological innovation in Japanese auto companies and the technology policies of the Japanese government. This chapter will provide valuable information for policymakers in Asian countries, many of which are currently experiencing rapid motorization and increasingly serious problems caused by societal dependence on automobiles.

Chapters 3 to 7 of this book are quantitative analyses of the impact of two types of government policy: fuel efficiency regulations and policies designed to promote the growth of intelligent transportation systems. Similar analyses of the situations in Europe and the US have already been published in several international journals and in a number of books, as mentioned above. However, to date no book written in English has discussed these points in the Japanese context. To our knowledge, Chapters 5 and 7 are the first study to consider the economic features of safety-related ITS technologies, a field in which Japan is playing a leading role, in the context of an investigation of the market diffusion mechanisms and economic welfare. These chapters discuss the economic features of various types of ITS – including the autonomous detection-type driving support systems, the roadside information-based driving support systems, and the inter-vehicle communication-type driving support systems – as well as policies to stimulate their market diffusion. We are confident that the analyses in this part of the book will be interesting reading, not only for researchers and policymakers in Asian countries but also for those in European countries.

Chapter 8 of this book deals with various problems related to automobile traffic, as well as the current state of automotive technology policies, in China. As in the case of the Japanese automotive industry, there are many books discussing the Chinese automotive industry from the production perspective, but to date no book has offered a comprehensive discussion of Chinese government technology policies. As China is certainly the most interesting market for the global automobile industry, as well as a site of significant automobile production, the analysis in this book will be a valuable resource for automobile industry personnel all over the world.

1.4 A brief synopsis of the chapters of this book

We will now briefly summarize each chapter in this book.

The current prosperity of the Japanese automobile industry was accomplished over the course of a mere 60 years, starting essentially from scratch at the end of World War II, and thus clearly demonstrates the world-class competitiveness of Japan's manufacturing industry. Therefore, we believe that reviewing the historical development of Japan's post-war automobile industry, and the government policies that guided that development, can provide important guidance for the formulation of future industrial policies in emerging countries. Chapter 2, 'Automotive Technology and Public Policy in Japan: A Historical Survey', by Masayuki Sano, Masanobu Kii, and Hiroaki Miyoshi, surveys technological innovations and technology policies in post-war Japan, then uses this survey as a backdrop for discussing the crafting of future automobile technology policies and the challenges attendant on this process. The authors observe that the automotive technologies required to address environmental and safety problems encompass not only core automobile technologies but also various other technologies. The authors also suggest that social systems, such as traffic and recycling systems, will be increasingly important in the future, and that effective use of information and communication technologies will grow ever more critical. In view of these trends, the conclusions of the chapter are (1) that governments should consider expanding the use of economic measures as an important addition to their toolkit of available policy options, and (2) that policy assessments based on quantitative analysis methods, such as cost–effect analysis, will be indispensable for building consensus among the parties affected by new policies.

Chapters 3 and 4 assess regulations on automotive technologies from the perspective of fuel efficiency, including an overview of future directions. Many developed countries today are making efforts to regulate vehicle fuel efficiency as a means of saving energy and combating global warming.

In Chapter 3, 'Costs and Benefits of Fuel Economy Regulations: A Comparative Analysis of Automotive Fuel Economy Standards and the Corresponding Benefits for Consumers', authors Masanobu Kii and Hiroaki Miyoshi compare two methods of fuel economy regulation – the weight-class-based method (currently employed in Japan) and the CAFE method (employed in the US) – to assess the impact of each method on the prices of vehicles and on the economic benefits offered to consumers. Using recent data on fuel-efficient technologies and their costs, the authors analyze the impact of the two regulatory structures on the prices of vehicles by considering the risks of noncompliance with CAFE standards against the volatility of the sales distribution, which is insignificant in the weight-class regulation. This analysis demonstrates that vehicle price increases due to the installation of fuel-efficient technologies are likely to be smaller under weight-class-based regulations than under CAFE regulations at the 2015 target level. On the other hand, the authors also conduct a sensitivity analysis, which demonstrates that target levels set to more stringent standards than the 2015 target level would make the CAFE method *more* reasonable to adopt than the weight-class approach, because the CAFE approach offers a greater variety of technological options – such as weight reduction technology – for improving fuel efficiency. Indeed, fuel-efficient technologies increase vehicle price but reduce fuel consumption. Finally, the authors analyze the impact of fuel economy regulations on the economic benefit realized by consumers, considering the case of weight-class-based regulations. They conclude that the net savings realized by users with the regulations in place are likely to be positive, in other words that the regulations will yield economic benefits for consumers.

Chapter 3 thus demonstrates that Japan's weight-class-based approach to fuel economy standards is an appropriate policy framework for today's world. But will such an approach – in which governments directly regulate fuel economy – remain the optimal mechanism for improving fuel economy in the future? In Chapter 4, 'Future Directions for Fuel Efficiency Policy: Evolving from Fuel-Efficiency Standards Toward Indirect Regulations', authors Hiroaki Miyoshi, Masayuki Sano, Masanobu Kii, and Yuko Akune pose this question and respond with a resounding *no.* A consideration of potential responses to the problem of global warming leads inescapably to the conclusion that direct government regulation of fuel efficiency – the primary policy mechanism that has been used to address global warming in the past – is a fundamentally inappropriate response to the critical problem of climate change. The chapter begins with a survey of the process used in the past to craft automotive regulatory policy in Japan; the authors note that regulatory standard levels in Japan have traditionally been calibrated to the base technological levels expected to have been attained by car-makers at the time the standards go into effect, thus ensuring that no car-maker fails to comply with the standards. The authors then discuss Japan's automotive tax system, which functioned for many years as a dedicated revenue source for the construction and maintenance of roads, and point out that several features of Japan's automotive tax structure – such as the relative magnitudes of basic taxes – are problematic, both from an environment-friendliness perspective and on the basis of the fundamental principles governing the assessment of public utility fees. The authors reach two main conclusions. First, if Japan continues its traditional strategy of direct government regulation of fuel efficiency, and if the process that has traditionally been used in Japan to set regulatory standards is preserved into the future, then a number of serious problems will arise, including not only a failure to achieve consensus within the automobile industry but also a potential degradation in the variety of choices available to consumers in the vehicle marketplace. Thus Japan must shift away from traditional direct regulation policies and instead toward indirect regulations (economic regulations), using tools such as tax and subsidy policy. Second, and on a related note, Japan's automotive tax system must be reformed in a way that emphasizes travel-phase taxes.

While Chapters 3 and 4 focus on analyzing fuel efficiency policies on vehicles, Chapters 5 to 7 consider intelligent transportation systems (ITSs) – technologies that use communication networks to provide information and enhance safety during road travel. Many goods and services in the IT industry, including the Internet, cell phones, computer software, CDs, DVDs, and video games, have a common feature known as *network externality*. The term describes an economy of scale on the demand side, in which the benefit that an individual derives from a good or service increases as the number of individuals consuming the same good or service increases. On the other hand, when we consider ITS technologies from the perspective of externality, we should pay attention to the following two features that other goods and services in the IT industry do not possess. First, traffic congestion and traffic accidents, which we seek to avoid through the use of ITS technologies, are typically phenomena caused by multiple vehicles. As a result, the existence of ITS users can serve to reduce the risk of congestion and accidents for *non-users* as well, and thus an ITS may exhibit features that benefit society as a whole – it may be a 'public good'. Second, when an ITS exhibits attributes of a public good, the value the system offers to its users is correctly assessed only after deducting its value as a public good. This leads to interdependency among users, a phenomenon quite distinct from the network externality mentioned above.

Chapter 5, 'Economics of Intelligent Transport Systems: Crafting Government Policy to Achieve Optimal Market Penetration', by Hiroaki Miyoshi and Masanobu Kii, qualitatively analyzes the benefits of several ITS technologies – taking into account the unique features of ITSs mentioned above – and discusses the government policies that may be required to stimulate the growth of the systems until the optimal market penetration rates are achieved. The chapter concludes with a discussion of the government policies required to achieve the optimal market penetration rates, and demonstrates that tax and subsidy policies, the attainment of critical mass, and the calibration of customer expectations – both by industry and by government – are all important elements.

Chapters 6 and 7 establish mathematical models of two real-world systems based on the theoretical framework established in Chapter 5, then use these models to conduct simulation analyses of market penetration rates and other quantities. Chapter 6, 'Optimal Market Penetration Rates of VICS', by Hiroaki Miyoshi and Masayoshi Tanishita, considers the Vehicle Information and Communication System (VICS). VICS is a Japanese system that provides traffic-related information, such as real-time reports on traffic congestion, traffic accidents, or link-travel-time, to drivers through an onboard car navigation system equipped with a map display. The service was first launched in metropolitan areas in 1996; its coverage area was gradually expanded, and since February 2003 it has been available throughout Japan. Chapter 6 begins by developing a simulation model in accordance with the theoretical concepts discussed in Chapter 5. The authors then use this model to calculate a variety of quantities related to the VICS, including the market penetration rate, the optimal penetration rate, the dynamic stability of the optimal penetration rate, and the magnitude of the taxes and/or subsidies required to realize this optimal penetration rate. Although the analysis is conducted under certain simplifying assumptions, the results suggest that it will be necessary to impose taxes on VICS units in order to realize the optimal penetration rate.

One of the most active areas of ITS research relates to vehicle safety, and especially the prevention of traffic accidents. Although a number of safety-related ITS technologies have been developed, the two types of system that have attracted the most recent attention are the *roadside-information-based* driving support systems (hereafter referred as to car-to-infrastructure systems) and the *inter-vehicle-communication-based* driving support systems (hereafter referred as to car-to-car systems). The car-to-car systems are designed to avoid accidents through inter-vehicle communication. An advantage of such a system is that it can be used everywhere, as long as all participating vehicles are equipped with the communication device; there is no need to install a roadside infrastructure. On the other hand, if either vehicle in a potential altercation is *not* equipped with the device, then the two vehicles cannot communicate with each other, and the accident prevention potential is lost. Car-to-infrastructure systems are based on information collected from vehicle detectors installed on roads. Thus, this type of system works only in locations where roadside devices are installed; on the other hand, the system can help to prevent accidents as long as either of the vehicles in a potential altercation is equipped with the device. In Chapter 7, 'Market Penetration of Safety-Related ITSs', authors Masanobu Kii and Hiroaki Miyoshi discuss the market penetration dynamics of these two types of system. The authors first develop a simulation model in accordance with the theoretical concepts discussed in Chapter 5, then use this model to derive two primary conclusions. First, the dual-mode system – an intermediate entity combining the best features of car-to-car system and car-to-infrastructure system – ensures that onboard devices will diffuse naturally throughout the market. Second, urban structure has a significant impact on the efficacy of safety-related ITSs. This suggests that urban structure issues must be critical inputs to the ITS policymaking process.

Finally, Chapter 8, 'Transport Problems and Policy Solutions in China', by Lin Sun, supplements the discussion of the first seven chapters by introducing the current state of environmental and traffic problems, and of automotive technological policies, in China. The size of China's new vehicle market has exceeded that of the Japanese market since 2005 and that of the US market since 2009. With the rapidly growing automobile market and the increased number of vehicles comes steadily increasing demand for automotive fuel, both gasoline and diesel. This increase in automotive fuel consumption has led to increased emissions of vehicular pollutants, which are the primary contributors to air pollution in both urban and rural areas. In short, all the negative externalities of automobile transportation that had previously been experienced by developed countries are now gradually becoming significant in China. The chapter begins with a brief overview of vehicle-related policies implemented in China to date, then discusses future trends in technology policies, based on China's observations of the experiences of developed countries in coping with the same problems. Particularly intriguing is the author's discussion of the unique difficulties China faces in setting regulatory standards and economic incentives. In brief, while China has a powerful interest in developing its own national brands, many of its most important technologies are currently controlled by multinational enterprises. Under such circumstances, the Chinese government is concerned that the implementation of high-level regulations, and the provision of economic incentives for cutting-edge technologies, may result in favorable treatment for multinational enterprises and corresponding limitations on the growth of domestic brands.

1.5 Topics for future research

Having surveyed the research questions asked and answered in this book, we will conclude this introductory chapter with a look at some of the problems our research group will be studying in the near future.

A first theme of ongoing work is the development of analytical methods that more realistically reflect the current state of technology. For example, in our analysis of fuel economy regulations in Chapter 2, we conducted literature searches to estimate the impact and the cost of future technologies, but this is a field in which the pace of technological advance is rapid, and several of the technologies that we anticipated for the future have already come into widespread use today. Similarly, in our analysis of safety-related ITS technologies in Chapter 7, we significantly oversimplified the distinction between the types of accidents that can be prevented by the two technologies we considered, and our consideration of market penetration processes incorporated these simplifying assumptions. The impact of safety-related ITS technologies will depend not only on the technical characteristics of the systems, but also on how they are used in practice. In order for the results of our analyses to find applications in the development of technology or technology policy for the automobile industry, we will need to conduct a more thorough investigation of the current state of technology.

A second research goal is the design of a simulation model capable of comprehensively assessing the impact of changes to the automotive regulatory systems and the automotive tax code. In Chapter 3 of this book we analyzed the impact of fuel efficiency standards on vehicle prices and on the benefits offered to consumers. However, this analysis entirely ignored the impact of vehicle price changes on consumer behavior, and specifically on consumer choices regarding purchasing and driving cars. Similarly, in Chapter 4, we presented the results of simulations conducted using a simulation model we developed – the 'Road and Environmental Policy Assessment Model' (Akune *et al.*, 2008) – but this model is not sufficiently sophisticated to capture the behavior of car-makers in developing low-fuel-consumption vehicles or improving the fuel efficiency of their existing vehicle fleet. Moreover, this analytical model was designed with the limited goal of assessing revisions to the automotive tax code, and is not equipped to simulate the impact of regulatory policies. In order to conduct realistic assessments of a variety of government policies in the future, we will need to develop a more comprehensive simulation model that improves on all of these points.

A third objective of our research group is to promote international research collaborations among real-world policymakers, academics, and industrial professionals in Japan and in a range of foreign countries. In particular, we feel it is essential to conduct a study similar to that reported in this book for the BRIC nations – Brazil, Russia, India, and China. For example, fuel economy regulations were first introduced in China in 2004, but these were only absolute upper limits on fuel consumption. China has not yet progressed to the point of enacting sales-averaged corporate fuel economy standards of the type common in Japan and the US, but it seems inevitable that the Chinese government will eventually need to consider regulatory frameworks and tax structures similar to those in Japan. Our ultimate goal is to contribute to ensuring that the developing nations of the world, beginning with China, carry out the motorization of their societies in a healthy and sustainable way; we believe an excellent way to help achieve this objective is to launch international research collaborations and to offer Japan's own historical experience, and the research we have conducted to date, as invaluable background and reference material for use in crafting future government policy.

A final objective of our ongoing research, as we look forward to the future of the automobile-empowered society, is to conduct policy analyses with a clear eye toward upcoming trends, including new types of vehicles that confound conventional paradigms – which have tended to emphasize the goal of *personal* mobility over all else – as well as new ways of using vehicles, including car sharing and carpooling. Indeed, the advanced nations of the world – as well as some developing nations – are now, or will at some point during this century become, rapidly aging societies; in societies that were once entirely dependent on automobiles, it may prove increasingly difficult to retain our individual mobility and preserve our right to move about freely. Moreover, environmental constraints such as global warming, and limitations on natural resources and the availability of cheap energy, may soon render traditional patterns of vehicle ownership and use obsolete. If people are to continue to enjoy the benefits of their beloved automobiles even as societies evolve and grapple with a variety of constraints, and if the deleterious effects of automobile use are to be kept under reasonable control, we will inevitably be forced to grope our way toward new types of vehicles and new ways of using them; it is our sincere hope that the methods of economic and policy analysis presented in this book will play a constructive role in designing these next-generation paradigms for the role of automobiles in society.

References

Akune, Y., Miyoshi, H., and Tanishita, M. (2008) 'Jidosha kanren zeisei to keizai kosei' ('Effect of Automobile Taxation System Revision in Japan'), in Miyoshi, H. and Tanishita, M. (eds), *Jidosha no gijyutu kakushin to keizai kousei: kigyo senryaku to kokyo seisaku* (*Technological Innovation in the Automotive Industry and Economic Welfare*), pp. 115–142, Tokyo: Hakuto Shobo Publishing Company.

Bekiaris, E. and Nakanishi, Y. J. (eds) (2004) *Economic Impacts of Intelligent Transportation Systems: Innovations and Case Studies* (*Research in Transportation Economics*), Amsterdam: Elsevier.

Black, W. R. and Nijkamp, P. (eds) (2002) *Social Change and Sustainable Transport*, Bloomington IN: Indiana University Press.

Ghosh, S. and Lee, T. S. (2010) *Intelligent Transportation Systems: Smart and Green Infrastructure Design*, Boca Raton, FL: CRC Press.

Greene, D. L., Jones, D. W., and Delucchi, M. A. (eds) (1997) *The Full Costs and Benefits of Transportation: Contributions to Theory, Method and Measurement*, Berlin: Springer.

Ministry of the Environment, Japan (2010) *Nihon no onshitsu kouka gasu hai-syuturyou* (*Total Greenhouse Gas Emissions in Japan*). Available at http://www.env.go.jp/earth/ondanka/ghg/index.html (accessed November 9 2010) (in Japanese).

National Research Council (2003) *Effectiveness and Impact of Corporate Average Fuel Economy (CAFE) Standards*, Washington, DC: Joseph Henry Press.

Ryan, L. and Turton, H. (2007) *Sustainable Automobile Transport: Shaping Climate Change Policy* (Esri Studies Series on the Environment), Cheltenham: Edward Elgar.

Schafer, A., Heywood, J. B., Jacoby, H. D., and Waitz, I. A. (2009) *Transportation in a Climate-Constrained World*, Cambridge, MA: MIT Press.

Sevigny, M. (1998) *Taxing Automobile Emissions for Pollution Control* (New Horizons in Environmental Economics), Cheltenham: Edward Elgar.

Small, K. and Verhoef, E. (2007) *Economics of Urban Transportation*, Abingdon: Routledge.

Sperling, D. and Gordon, D. (2010) *Two Billion Cars: Driving Toward Sustainability*, Oxford: Oxford University Press.

Stough, R. (ed.) (2001) *Intelligent Transport Systems: Cases and Policies*, Cheltenham: Edward Elgar.

Womack, J. P., Jones, D. T. and Roos, D. (1990) *The Machine That Changed the World*, New York: Rawson Associates.

2

Automotive Technology and Public Policy in Japan: A Historical Survey

Masayuki Sano, Masanobu Kii, and Hiroaki Miyoshi

2.1 Introduction

The automobile population in Japan has risen to approximately 80 million vehicles and is the world's second largest, trailing only that of the US. Japan is also second to the US in the number of vehicles produced, while it leads the world in the production of passenger cars. The Japanese automobile industry is thus not only a key industry within Japan, but also commands an important position in the world's industrial structure.

The current prosperity of the Japanese automobile industry was accomplished over the course of a mere 60 years, starting essentially from scratch at the end of World War II, and thus clearly demonstrates the world-class competitiveness of Japan's manufacturing industry. Therefore, we believe that reviewing the historical development of Japan's post-war automobile industry, and the government policies that helped to shape it, can provide important guidance for the formulation of future industrial policies in emerging countries.

This chapter examines the historical relationship between automotive technology policies and technological innovations in Japan, with an eye toward the design of policies for the future. We begin in Section 2.2 with a survey of technological innovations and technology policies in post-war Japan. In Section 2.3 we review future directions for technological innovations in automobiles. Finally, in Section 2.4, we discuss how future automobile technology policies should be crafted and the challenges attendant on this process.

2.2 A brief history of post-war automotive technology and technology policies

As is clear from Figure 2.1, the record of technical innovations emerging from Japan's post-war automobile industry was a result of steady corporate investment in research and development. But to understand the unique form in which the Japanese automobile industry developed, we must first understand the *social* background that gave rise to the government policies that in turn enabled such massive corporate investments.

In this section, we classify into six categories the technologies that were needed in each era of Japan's post-war economic growth: 1) mass production technologies, 2) safety technologies, 3) technologies for combating air pollution, 4) technologies for energy conservation and environmental protection, 5) technologies to further the goal of a sustainable society, and 6) automotive information and communication technologies. This classification then enables a comprehensive examination of the relationship between Japanese technological innovations

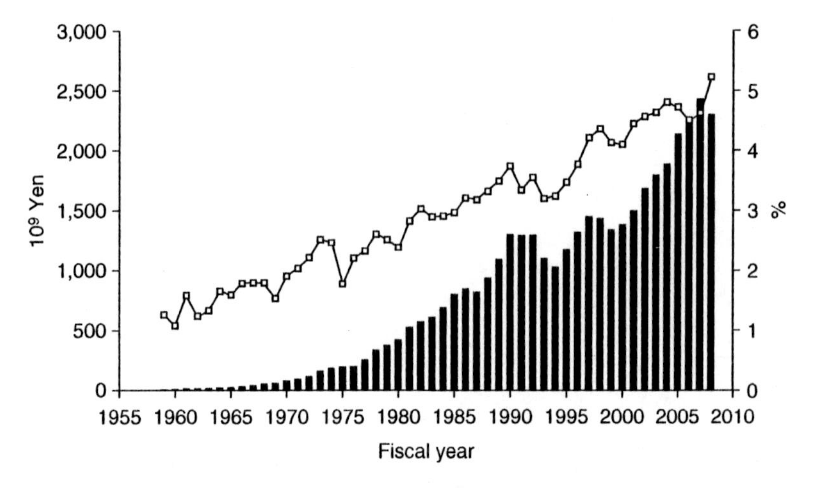

Figure 2.1 History of intramural R&D expenditures in the Japanese automobile industry

Note: Bars indicate the automotive industry's intramural R&D expenditures (with units given on the left vertical axis). The curve indicates the automotive industry's R&D expenditures as a percentage of sales (with units given on the right vertical axis).

Source: Prepared by the authors based on the *Survey of Research and Development* from the Statistics Bureau of Japan's Ministry of Internal Affairs and Communications.

and government policies related to automobiles, considered against the backdrop of the social climate prevailing in each era.

2.2.1 Before the establishment of mass production technologies

The Japanese automobile industry suffered great damage in World War II, with particularly catastrophic damage to the passenger-car sector. In the period between the defeat and 1949, the production of passenger cars was prohibited, or restricted, by the General Headquarters of the Allied Forces (GHQ). Post-war automobile production was limited to production of about 20 thousand trucks per year, with the permission of the occupying force, and Japan depended on foreign imports for passenger cars.

When Japan regained independence in 1952, the Ministry of International Trade and Industry (today's Ministry of Economy, Trade and Industry) imposed restrictions on foreign investment and limited imports of assembled cars, including used cars. Under these protectionist trade policies, the Japanese automobile industry was able to develop without direct threat of foreign capital investment or vehicle imports. In addition, in 1951, the Road Transport Vehicle Act was enacted. This law established a 'Safety Standard for Road Transport Vehicles' (referred to below as the 'Safety Standard') and launched the Automobile Inspection and Registration System (referred to below as the 'Inspection System').

In the 1960s, in addition to making steady efforts in product development, car-makers extensively advertised the reliability and performance of their products in an effort to compete with other companies. The Installment Sales Act was established in 1961, making passenger cars affordable for middle-income consumers.

By the end of the 1960s, rapid economic growth had led to a dramatic increase in car sales. The number of passenger cars produced exceeded 1 million by 1967, and the total car population grew to nearly 7 million by the end of 1969 (see Figure 2.2). Among the 61 types of passenger car listed in *240 Japanese Automotive Technologies* (Society of Automotive Engineers of Japan, Inc., 2007), 35 were introduced in the 1960s, and it is fair to say that the 1960s was the era in which Japan's automobile industry consolidated its technological base and established its mechanisms for the mass production of passenger cars. It should be also noted that 1960 was the year in which Toyota introduced its 'Kanban System', a just-in-time production system that has significantly influenced the development of mass production technologies.

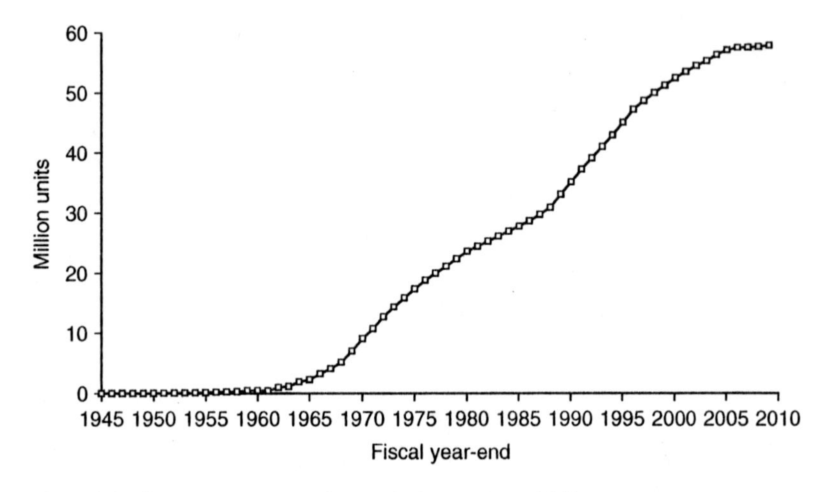

Figure 2.2 Passenger-car population in Japan since 1945

Source: Prepared by the authors based on ultra-long-term statistical data found in the 2010 *EDMC Handbook of Energy & Economic Statistics in Japan*, edited by The Institute of Energy Economics, Japan.

2.2.2 Safety measures

By 1971, the passenger vehicle population in Japan had grown to 10 million (see Figure 2.2). In the years since this milestone, the increase in traffic accidents due to rapid motorization and the increase in air pollution due to vehicle emissions have emerged as important social problems, and measures to promote safety and reduce air pollution have become important challenges for the Japanese government and automobile industry. We begin our consideration of these issues by surveying the history of vehicle safety measures in Japan.

Japanese automotive safety measures may be classified into two categories. One set of measures are those falling within the framework established by the Road Transport Vehicle Act, which established a 'Safety Standard' for items such as vehicle body structure, on-board equipment, maximum passenger capacity, and maximum cargo capacity, with compliance assessed at mandatory periodic automobile inspections. The second category comprises measures based on the Road Traffic Act, which attempts to promote safety in vehicle usage through traffic rules and the licensing system. To ensure the protection of car users, a recall system was established in 1969, while a recall admonishment system, with penalties, was instituted in 1999 under the Road Transport

Vehicle Act, completing the system in place today. In addition, a 'Vehicle Assessment System', spearheaded by the Ministry of Transport (today's Ministry of Land, Infrastructure, Transport and Tourism), was created in 1995, and since that year the safety level of every car sold in the Japanese market has been assessed and published annually.

The Safety Standard has been amended and expanded more than 100 times since 1951. In the US, criticism of car-makers by consumer groups[1] led to the enactment in 1966 of the National Traffic and Motor Vehicle Safety Act. This legislation established rules for automobile design and established an administrative authority, the National Highway Safety Bureau (NHSB) (the predecessor of today's National Highway Traffic and Safety Administration (NHTSA)), within the US Department of Transportation. The system imposed specific regulations based on Federal Motor Vehicle Safety Standards (FMVSS) set by the NHSB. This development in the US motivated the introduction in Japan of safety measures for heavy-duty vehicles, including mandatory installation of rear under-run protectors, under-run protectors, and fail-safe brakes. These provisions were introduced into Japan's Safety Standard in 1967.

Mandatory installation of seatbelts began in 1968, and measures relating to the body structure of heavy-duty vehicles, brake function, and headlights were introduced and strengthened throughout the 1970s and 1980s. In the 1990s, motivated by the stubborn refusal of the traffic fatality rate to fall below 10,000 a year, efforts were made to strengthen standards aimed at promoting safety at the time of collision. In the first half of the 1990s, the installation of anti-lock braking systems (ABS) was made mandatory, first for heavy-duty vehicles and later for medium-duty vehicles. Then, starting in 1993 with the introduction of standards for passenger protection in the event of a frontal collision, and mandatory full-scale crash testing for passenger cars, the Safety Standard has been gradually strengthened; standards for passenger protection in the event of a side collision, and in the event of an offset collision, were added in 1999 and in 2000, respectively. We should also note that the specifications for mini-sized vehicles (Japan's K-cars), although not part of the Safety Standard, were amended in 1998 to enlarge the size of the vehicles, in an effort to improve collision safety. Although airbags are not explicitly required by the Safety Standard, their installation is required in practice for compliance with passenger protection provisions. Important revisions to the Safety Standard since 2000 include the mandatory installation of speed limiters (enforcing a 90 km/h limit) in heavy-duty trucks and the introduction of pedestrian protection regulations. The first of these was imposed on new vehicles starting in 2003, and was subsequently expanded to cover existing vehicles by 2006. The second provision aimed to limit the shock delivered by a collision to a pedestrian's head, and was first implemented in 2005 through collision testing with crash-test dummies.

The Road Traffic Act includes seatbelt and child seat regulations. A non-binding provision requiring drivers and front passengers to wear seatbelts was introduced in 1971, and was gradually strengthened in the following years. This provision was made mandatory in 1986, and the most recent revision of the law, in 2007, mandated seatbelt use for rear passengers as well. As for child seats, a binding provision requiring child seats for children six years old and under has been in place since 2000.

Faced with these safety-related regulations, Japanese vehicle manufacturers have taken the Safety Standard as the minimum standard to be met, and have continued to invest steadily in developing technologies to assure compliance. In addition, as increasing consumer interest in automotive safety enhances the market value of safety features, car-makers have opted voluntarily to provide extended safety-related technologies beyond those mandated by the Safety Standard, particularly for passenger cars. High-end, ordinary-sized passenger cars today routinely offer not only passive safety technologies, such as various types of airbags, but also options for active technologies based on information and communication services, such as the Collision Avoidance System; such options are also gradually expanding to small- and mini-sized mid-market passenger cars.

2.2.3 Measures for addressing air pollution

The problem of pollution first emerged in the late 1960s as an unpleasant side effect of rapid economic growth, with air pollution caused by car emissions joining pollution by the heavy-chemical and electricity-generation industries as serious societal problems. Since then, the design of measures to address vehicle-related air pollution has been an important challenge for Japan's government and automobile industry, and the process has had significant influence on the post-war development of automotive technologies.

a) Introduction of environmental regulations and measures for unleaded fuels.

The first regulations on automobile emissions were a set of instructions issued by the Ministry of Transport (today's Ministry of Land, Infrastructure, Transport and Tourism) in 1966,[2] which imposed regulations on carbon monoxide (CO) emissions. In 1968, the Air Pollution

Control Act was established, under the Basic Act for Environmental Pollution Control approved the previous year, and tolerances on CO emissions per vehicle were imposed. The Noise Regulation Act, established the same year, set limits on automotive noise emission.

The Ushigome-Yanagicho lead pollution problem of 1977, and the Suginami photochemical smog problem which followed it, led to increasing public concern over automotive pollution. Meanwhile, also in 1977, the US Congress passed a revised version of the Clean Air Act (Muskie Act). This legislation inspired Japan's automobile industry, for which the US was a major export market, to invest research expenses amounting to approximately 2.5 per cent of sales into technology development (see Figure 2.1), and thus represents an epochal turning point in the history of Japan's post-war automotive technology development. As mass-produced vehicles began to be sampled to assess compliance with specific automotive emissions regulations, vehicle design accuracy improved, which in turn contributed to improvements in overall automobile quality.

The years 1971 to 1979 saw the development of a number of new technologies, including Honda's Compound Vortex Controlled Combustion (CVCC) engine – the world's first engine to comply with the Muskie Act – as well as the Electronically Controlled Gasoline Injection (ECGI) system for accurate combustion control, the lean-burn method, and the three-way catalyst for simultaneous detoxification of NO_x, HC, and CO. These innovations formed the technological basis for subsequent methods of reducing and purifying emissions from gasoline-powered engines. Although such technologies are each independently effective, the greatest effect is achieved when a variety of processes, ranging from combustion control to post-treatment, are *combined*, and this integration of technologies has contributed significantly to improving the design of automobiles from the perspective of energy usage and environmental protection. The 1970s also witnessed dramatic improvements in techniques for measuring vehicle emissions, and the basis of the Inspection System for regulating vehicle emissions was consolidated during this period.

On the other hand, although actions taken by the Japanese government lagged behind corresponding developments in the US, several additions were made in 1971 to the list of substances controlled by automobile exhaust-gas regulations, including nitrogen oxides (NO_x), total hydrocarbons (THC), particulate matter (PM), and lead (Pb). In 1973, the Ministry of the Environment announced its '1973 Emission Control Standards', limiting NO_x and THC emissions. This was the beginning of major automobile exhaust-gas regulations in Japan, and regulatory standards have been strengthened almost every year since, culminating in today's standards. Limits on NO_x from gasoline-powered passenger cars have been steadily tightened, becoming the world's most stringent with the '1978 Emission Control Standards'. By that time, all Japanese car-makers were able to produce cars meeting environmental regulations, but no US manufacturers could produce cars complying with the Muskie Act, whereupon the implementation of the Act was postponed. It was only in the 1990s that US manufacturers became able to produce vehicles meeting the levels mandated by Japanese law in 1978.

Although the addition of lead to gasoline can prevent reciprocal engine knocking and improve octane number, lead is not only toxic but also degrades the performance of vehicle emissions control systems (catalysts). As this poses severe challenges for the reduction and purification of vehicle emissions, car-makers and the oil industry cooperated in their efforts to promote unleaded fuels. As a result, all cars produced after 1972 were made compatible with unleaded gasoline, while regular gasoline became completely unleaded in 1975. Lead continued to be added to premium gasoline after 1975, but proportions had decreased dramatically by around 1980, and in 1987 Japan became the world's first nation in which all types of gasoline were unleaded. The US followed suit in 1996, and the countries of the EU at the end of the 1990s. As of 2009, only 11 countries continue to use leaded gasoline.

b) Introduction of vehicle-type restrictions.

In the 1980s, regulations on automobile emissions expanded to encompass NO_x and PM emitted from diesel-powered trucks and buses. Specific regulations on diesel vehicles were introduced in 1972 for black exhaust and in 1974 for CO, THC, and NO_x, and these standards were steadily tightened in the following years, with particular focus on NO_x, but the critical turning point came with the 1989 proposal, by the Central Council for Environment Pollution Control, of short-term and long-term targets for emissions reductions. The proposal included targets and timelines for reducing NO_x from gasoline-powered medium- and heavy-duty vehicles, for which regulation was overdue, as well as for reducing NO_x and PM emissions from diesel-powered vehicles. For diesel-powered trucks and buses, the short-term target (to be achieved during 1993–94) and the long-term target (to be achieved during 1997–99) were known as the short-term diesel regulation and the long-term diesel regulation, and the regulatory standards have been tightened in the years since, in accordance with the original proposal.

However, despite the tightening of individual emissions regulations, the air-pollution situation in urban areas, especially areas along heavily traveled roads, failed to improve, and environmental standards for nitrogen dioxide (NO_2) and suspended particulate matter (SPM) in the air continued to be violated (see Figure 2.3). Therefore, in 1992, the Act Concerning Special Measures for Total Emission Reduction of Nitrogen Oxides from Automobiles in Specified Areas (the Automotive NO_x Law) was enacted. This was a type of regulation known as a 'vehicle-type restriction'. It defined certain target areas, within which the new registration, transfer of registration, or renewal of registration of any vehicle which failed to meet emission standards for that target area was prohibited. One hundred and ninety-six localities in the Tokyo metropolitan area and the Osaka and Hyogo regions were designated as target areas. The legislation had the effect of encouraging drivers to replace old vehicles, and placed additional burdens on vehicle users in the cargo-transport and other industries. However, as the legislation did not require the development of new automotive technologies, it did not spur new technology development.

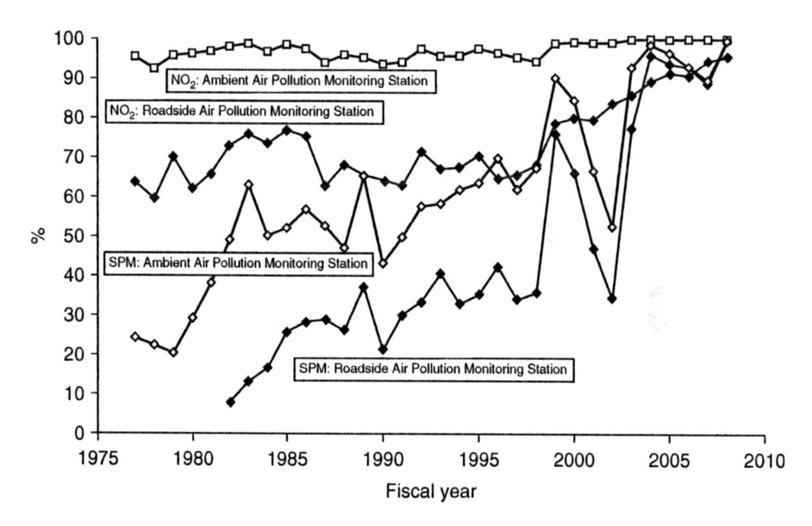

Figure 2.3 Compliance with environmental standards for NO_2 and SPM as measured by Air Pollution Monitoring Stations

Source: Prepared by the authors based on the Ministry of the Environment's 'Annual Report on the Environment in Japan' (1978–2006), 'Annual Report on the Environment and the Sound Material-Cycle Society in Japan' (2007–08) and 'Annual Report on the Environment, the Sound Material-Cycle Society and the Biodiversity in Japan' (2009).

The Automotive NO_x Law was amended in 2001, as its enactment had not improved NO_2 -related environmental conditions in the target areas, and the severity of SPM-related air pollution had heightened concerns over the health effects of SPM emissions from diesel vehicles. In the amended version, SPM joined NO_x as a target of the regulations, Aichi and Mie prefectures were added to the list of regions concerned, and 276 municipalities were designated as target areas. The amended law is known as the 'Automotive NO_x/PM Law.'

c) Measures addressing diesel vehicles and fuel modifications.

Even with the new vehicle-type restrictions, the air in urban areas continued to violate environmental standards for NO_2 and SPM (see Figure 2.3). As a result, more drastic emission-control measures were required, and individual regulations on vehicle emissions were further tightened.

For gasoline-powered vehicles, the second recommendations of the Central Environment Council (CEC) regarding short-term regulations were proposed in 1997 and implemented in 2000 (in 2002 for mini-sized passenger cars). In addition, a new long-term regulation was proposed, namely that NO_x emissions be further reduced by 50 to 70 per cent by 2005. For diesel vehicles, the third CEC recommendations in 1998 proposed new short-term regulations, which were implemented in 2002, although they came into force only in 2004 for heavy-duty vehicles such as trucks and buses. Likewise, as a new long-term regulation, the council proposed that NO_x emissions be further reduced by 41 to 50 per cent by 2007. PM standards were also tightened in tandem with measures to combat NO_x.

Faced with this trend toward ever-tighter standards, car-makers, and especially diesel-vehicle manufacturers, focused their efforts on developing technologies to meet regulations, with particular emphasis on engine-body-related measures such as adjustments to fuel injection timing, improvements to the combustion chamber and inlet–outlet port system, and higher fuel-injection pressure. However, among the emission-reduction technologies developed in the 1980s and 1990s, only the common-rail fuel injection technology has turned out to be significant – a situation to be contrasted with the progress in gasoline engines around the time of the Muskie Act, when important new technologies were developed in rapid succession. The difference arises because engine-related technologies and emission post-treatment technologies alone do not suffice to bring about significant reductions of NO_x and PM emissions in diesel vehicles; instead, it is also necessary to decrease the sulfur content of the diesel fuel oil as much as possible. In fact, although the sulfur content of diesel oil was regulated at 2,000 ppm in 1993 and at 500 ppm in 1997, the fourth CEC recommendation in 2000 tightened the schedule for long-term diesel-vehicle regulations to 2005 and proposed to limit the sulfur content of diesel oil to 50 ppm.

As explained above, measures to meet new long-term regulations for diesel vehicles required huge capital investments by the oil industry in addition to the technology development efforts of the automotive industry. For this reason, the Japanese automobile and oil industries conducted a joint research program known as JCAP (Japan Clean Air Program, 1997–2001), following the example of similar programs conducted by the automotive and oil industries in the US and Europe. The program included joint technology R&D activities as well as attempts to adjust the relative share of the burden shouldered by each industry. The result was a realistic roadmap toward the provision of diesel oil with sulfur content of 50 ppm by 2005 and compliance with long-term diesel regulations.

This roadmap allowed the establishment of specific regulatory values for new long-term regulations in the fifth CEC recommendation in 2002. In addition, further reductions in fuel sulfur content were proposed, and the seventh CEC recommendation in 2003 suggested limits of 10 ppm for both gasoline and diesel fuel. The automobile industry and the oil industry established JCAP II (2002–06) to succeed JCAP I, with continuing focus on reducing fuel sulfur content. As a result, diesel fuel with sulfur content of less than 10 ppm became available in 2005.

d) Post-new-long-term regulations.

Next, the eighth CEC recommendation in 2005, in seeking to develop a framework for post-new-long-term regulations for diesel vehicles, noted the introduction of diesel particulate filters (DPFs) and of diesel fuel with sulfur content of 10 ppm or less, suggested measures such as the introduction of NO_x post-processing apparatus, and suggested that exhaust-gas regulations for diesel vehicles could, in essence, be made equally stringent as those for gasoline vehicles, both for PM and for NO_x. With regard to PM in particular, the development of DPF technology has made it possible to pursue 'PM-free' vehicles, in which the PM content of exhaust gas is lower than the smallest content measurable with current measurement methods.

These recommendations were incorporated into regulations adopted in 2009 (the 'post-new-long-term' regulations), which regulated NO_x content of diesel vehicles at 40–65 per cent of the levels mandated by the new long-term regulations and PM content at 53–64 per cent of the levels mandated by the new long-term regulations. In addition, a PM standard equivalent to that applied to diesel vehicles (0.005 g/km) was mandated for some gasoline vehicles (namely, those equipped with direct cylinder fuel injection engines containing absorption-type NO_x reduction catalysts). Based on these post-new-long-term regulations, and premised on the availability of gasoline and diesel fuel with sulfur content of 10 ppm or less, car-makers today are proceeding with the development of technology to counteract environmental pollution. For diesel vehicles in particular, a number of vehicles incorporating new technologies are starting to come to market, including DPFs combined with accurate electronic control of engines, urea selective catalytic reduction (Urea SCR) systems, and diesel particulate–NO_x reduction (DPNR) catalysts.

e) Automobile green tax and 'Eco-car' subsidies.

Automotive technology policies to combat air pollution have traditionally focused on regulations and vehicle-type restrictions. Two economic policies have been introduced to date: 'Automobile Green Tax' and 'Eco-car' subsidies. The 'Automobile Green Tax,' which reduces the automobile tax and the automobile acquisition tax on low-emission vehicles, was introduced in 1975. 'Eco-car' subsidies, designed to encourage drivers to trade in older vehicles for more environmentally-friendly 'eco-cars', were introduced in 2009. The detailed explanation of these two plans is covered in depth in Chapter 4.

2.2.4 Measures to promote energy conservation and combat global warming

The first oil crisis of 1973 plunged the world into economic turmoil and had a significant impact on the Japanese economy, which had been almost entirely dependent on the Middle East for oil imports. The crisis spurred the development of energy conservation technologies in Japan, and an 'Act on the Rational Use of Energy' (referred to below as the 'Energy-conservation Law') was enacted in 1976, aiming to promote significant energy conservation in all sectors of the economy, including transportation, construction, and manufacturing. When substantial improvements in energy efficiency became an important common challenge across the developed world after the second oil crisis of 1976, specific energy-saving objectives were introduced for automobiles. Gasoline-powered passenger cars were designated as a particular target of the Energy-conservation Law, and fuel economy standards were established.

As mentioned in Section 2.2.3, car-makers at that time were already concentrating on the development of technologies to combat air pollution. The achievement of fuel economy standards added a further challenge, and vehicle manufacturers were obliged simultaneously to advance air-pollution countermeasures and safety features without increasing vehicle weight, which would have degraded fuel economy. The combined challenges placed severe constraints on technology development.

While air-pollution countermeasures have progressed largely on the basis of a relatively small number of core technological innovations (such as the three-way catalyst), no such definitive core technology has emerged to improve fuel economy; instead, the challenge requires an aggregate of multiple basic technologies. Such elemental technologies include lighter car bodies, improved engine efficiency (improved heat efficiency and reduced friction loss), reduced power-train losses (improved transmission efficiency), reduced running resistance (air resistance and rolling friction), and improved efficiency of accessories (such as air conditioning, power steering, and idling stop devices). Realizing lighter vehicle bodies requires weight reductions, by the gram, in every single component of the vehicle. Fuel economy requirements in the 1980s triggered intense development efforts by car-makers, and, as shown in Figure 2.1, investment in research remained steady at approximately 3 per cent of the sales volume throughout this period.

A glance at *240 Japanese Automotive Technologies* reveals that fuel economy technologies began to appear in the 1980s, including lean-burning engines, direct cylinder-injection of gasoline, the miller cycle engine, and the Continuously Variable Transmission (CVT). In addition, Japanese car-makers are good at combining many core technologies in the product development process, and this has been one of the key factors underlying the international competitiveness of the Japanese automobile industry, which was said to be the world's strongest in the 1980s.

Global warming emerged as a significant problem in the second half of the 1980s. With the adoption in 1992 of the 'United Nations Framework Convention on Climate Change' (UNFCCC), fuel economy requirements for automobiles began to draw attention as a means of combating global warming. Fuel economy standards for gasoline-powered passenger cars were amended in 1993, with a target year of 2000. Similarly, new fuel economy standards for gasoline-powered trucks were established in 1996, with a target year of 2003. After target values for greenhouse gas emissions in developed countries were set by the Kyoto Protocol in 1997, the Energy-conservation Law was significantly amended in 1998, and various new measures were adopted, including the introduction of the 'top-runner' approach to automobile fuel economy standards, energy-saving standards for electronic devices, and mandates for large-scale energy-consuming factories to prepare mid- and long-term energy-saving plans.

Since this amendment to the Energy-conservation Law, the 'top-runner' approach has been applied to automobile fuel economy standards, and these standards have been strengthened for a gradually expanding number of target vehicle types. Fuel economy standards for passenger cars (both gasoline and diesel) and for light-duty trucks were first set in 1999. The addition in 2003 of fuel economy standards for liquefied petroleum gas (LPG) passenger cars then ensured that standards existed for all passenger vehicles. In 2006, Japan became the first nation in the world to impose fuel economy standards on medium- and heavy-duty vehicles (such as trucks and buses). In 2007, the fuel economy standards for passenger cars and light-duty trucks were revised and tightened, and a standard for light-duty buses was also introduced. Fuel economy standards are thus in place today for all passenger cars and trucks, as summarized in Table 2.1.

These Japanese fuel economy standards are set separately for each vehicle weight class, and each car-maker is required to ensure that the average fuel economy of the vehicles it ships within a given weight class (for passenger cars, for example, there are 16 weight classes) does not fall below the target value for that weight class. This method differs from the Corporate Average Fuel Economy requirements (CAFE), which were introduced in the US in 1975 in response to the first oil crisis. In CAFE, the relevant quantity is the overall average fuel economy of all vehicles shipped by a car-maker, and each car-maker can thus use marketing strategies to increase the relative proportion of sales of light vehicles (which have better fuel efficiency) to meet the standard. However, as fuel economy standards in Japan set separate targets for each vehicle weight class, Japanese car-makers cannot utilize such a strategy.

In the second half of the 1990s, car-makers began intensifying their efforts to develop technologies to improve fuel economy, motivated by fuel economy requirements and by international market trends demanding high-performance fuel-efficient cars. In addition, car-makers undertook a long-term, globally minded commitment to develop technologies for alternative-energy vehicles with reduced CO_2 emissions, such as natural gas vehicles (NGV) and fuel-cell hybrid cars, as well as technologies for alternative fuels such as bioethanol. The start of commercial production and sales of hybrid passenger cars (the Toyota Prius) in 1997 represented the market debut of the fruits of these efforts.

Table 2.1 Overview of automotive fuel economy standards

Date	Item	Target fiscal year
June 1976	Enactment of Act on the Rational Use of Energy (Energy-conservation Law)	
December 1979	Fuel economy standards for gasoline-powered passenger cars	1985
January 1993	Fuel economy standards for gasoline-powered passenger cars	2000
March 1996	Fuel economy standards for gasoline-powered trucks	2003
June 1998	Amendments to Energy-conservation Law; introduction of top-runner standards	
March 1999	Fuel economy standards by top-runner approach for passenger cars and light-duty trucks	2010 (gasoline cars) 2005 (diesel cars)
July 2003	Fuel economy standards by top-runner approach for LPG passenger cars	2010
March 2006	Fuel economy standards by top-runner approach for medium/heavy-duty vehicles (such as trucks and buses)	2015
July 2007	Fuel economy standards by top-runner approach for passenger cars, light-duty buses and light-duty trucks	2015

Source: Prepared by the authors on the basis of Japanese legislation.

As a result of these efforts by car-makers, fuel economy for gasoline-powered passenger cars has improved significantly, and all vehicle manufacturers had already met 2010 target values by 2007, several years ahead of schedule. On the other hand, fuel economy for trucks saw little improvement in the 1990s, as many of these vehicles use diesel engines, and the development of emissions-reduction technologies was considered a more urgent priority. However, now that car-makers enjoy realistic prospects of developing technologies to reduce vehicle emissions below the limits set by 'post-long-term' regulations, it seems likely that diesel-vehicle fuel efficiency will improve toward compliance with the fuel economy standards for heavy-duty vehicles established in 2006 (for which the target year is 2015).

2.2.5 Measures to achieve a sustainable society

Combating the problem of global warming, which emerged in the 1990s, also requires that end-of-life vehicles (ELVs) be recycled. Parts and metals can be collected from ELVs, and, by the 1970s, a system had been established in which approximately 80 per cent (by weight) of ELVs were recycled by local auto dismantlers and other operations, while the remainder was deposited in landfills as automotive shredder residue (ASR). However, in the 1980s, due to (1) elevated costs for disposing of ASR due to a shortage of industrial waste dump sites and (2) declining prices for scrap steel, it became costly to recycle ELVs, and increased illegal dumping by traders wishing to avoid disposal costs became a social problem. In a 1990 illegal dumping case, in Teshima Island in Japan's Kagawa prefecture, some 560,000 tons of industrial waste, including much automobile waste such as ASR and used tires, was illegally dumped or disposed of (openly burned).

This situation made it strongly desirable in the 1980s to establish a stable system for recycling ELVs. In the meantime, new challenges had emerged, including the collection and destruction of chlorofluorocarbon chemicals (CFCs) and the appropriate disposal of explosive airbags. To address these issues, a collaboration between the Japanese government, the automobile industry, and the recycling and waste-management industries gave rise to an 'Act on Recycling, etc. of End-of-Life Vehicles' (known as the Automobile Recycling Law), aimed at mandating appropriate recycling and disposal practices for ELVs. The act was enacted in 2002 and went into effect in 2005.

The act required that car-makers reclaim, recycle, and dispose appropriately of ASR, CFCs, airbags, and other disposable components from ELVs, and that consumers cover related costs at the time of new vehicle purchase. Vehicle users who purchased their vehicles before the implementation of the act were required to cover costs prior to their first automobile inspection after the enactment of the law. The act further designated a set of electronic information and communication technologies through which the recycling and disposal systems would be comprehensively controlled.

The provision of the Automobile Recycling Law that will be most difficult to achieve is the progression of targets it sets for ASR recycling rates: over 30 per cent by 2005, over 50 per cent by 2010, and over 70 per cent by 2015. In order to achieve these targets, the automobile industry divided domestic and foreign car-makers into two teams[3] to concentrate efforts. The idea underlying this two-team scheme was that competition would spur the development of an optimal recycling system, the cost of which would be shouldered by vehicle users, but that car-makers would nonetheless need to collaborate with one another on initiatives to improve the operating efficiency of the system.

2.2.6 Measures for networking, globalization, and introduction of information technologies

Information and communication technologies have been a mainstay of automotive design and production for many years. Japanese car-makers had introduced computer-aided design (CAD) system into automobile design as early as the 1970s, and information and communication technologies have been incorporated into automobiles from the earliest stages.

Information and communication technologies have also taken root as core technologies for vehicle bodies. Core technologies for addressing automobile emissions became common in the 1970s, including engine control technologies such as electronic fuel injection control, microprocessors running embedded programs, power-train control techniques (such as automatic transmission (AT) and CVT), measuring devices, electronic control of installed equipment (such as air conditioners and audio devices), and vehicle body controls (ABS, power steering, and power windows).

The rapid development of information and communication technologies in the 1990s created increasing demand for technologies to address traffic problems and to improve the safety and efficiency of automobile travel. The concept of intelligent transportation systems (ITSs) emerged during this period, and ITS research has advanced significantly in the ensuing years.

The expansion of Japan's communication infrastructure enabled rapid growth in Internet usage at the end of the 1990s. In 2001, an IT Strategy Headquarters division was founded within Japan's Cabinet, with ITS taken as a specific example of a societal image for which to strive and positioned as a key component of Japan's IT strategy. ITS policies are currently under joint administration by the Ministry of Land, Infrastructure, Transport and Tourism, the Ministry of Economy, Trade and Industry, the Ministry of the Environment, and the National Police Agency. Table 2.2 lists nine sectors in which ITS services and systems have been developed. Of these, systems for the provision of traffic information, such as VICS, have enjoyed particularly rapid growth.

Among the services and systems listed in Table 2.2, vehicle safety is perhaps the most relevant to the development of automotive technologies, and Japanese car-makers have demonstrated their commitment to this challenge. Meanwhile, improved navigation systems and optimized traffic management are expected to have positive effects on travel efficiency and environmental protection, and these areas have also witnessed continuous advancement in services and technologies.

Table 2.2 Development sectors for ITSs in Japan

	Sector	Example of user services
1	Navigation systems	Vehicle Information and Communication System (VICS)
2	Electronic toll collection systems	Electronic Toll Collection System (ETC)
3	Vehicle Safety	Advanced Cruise-Assist Highway Systems (AHS), Advanced Safety Vehicle (ASV)
4	Optimization of traffic management	Optimization of traffic flow; provision of traffic restriction information for incident management
5	Road management efficiency	Management of special permitted commercial vehicles; provision of roadway hazard information
6	Public transportation	Provision of public transportation information; assistance for public transportation and operations management
7	Commercial vehicle operating efficiency	Assistance for commercial vehicle operations management
8	Pedestrians	Pedestrian route guidance; vehicle–pedestrian accident avoidance
9	Emergency vehicles	Automatic emergency notification; route guidance for emergency vehicles; support for relief activities

Source: Ministry of Land, Infrastructure, Transport and Tourism Website.

2.3 Future directions for innovations in automotive technology

In Section 2.2, we considered the development of automotive technologies in post-war Japan, from the early establishment of a technological base, to the implementation of mandatory safety measures, through the consolidation of technologies for combating air pollution and protecting the global environment, and up to the information and communication challenges of today. In this section, we investigate future directions and policy challenges in automotive technologies, with special focus on environmental and energy technologies and on safety-related ITSs.

2.3.1 Environmental and energy technologies

a) General trends.

The Society of Automotive Engineers of Japan (2007) classified environmental and energy technologies into four categories: technologies to prevent global warming, technologies to reduce air pollution, technologies to reduce automotive noise, and recycling technologies.

Fuel economy requirements have been promoted as a means of ameliorating global warming, and their effect has begun to appear in the form of a reduction in total CO_2 emissions from automobiles. The Society of Automotive Engineers of Japan (2007) proposed the realization of 'two-liter cars' (automobiles able to travel 100 km on the energy equivalent of two liters of gasoline) and suggested, as strategies for achieving this goal, (1) promoting the development of fuel-efficient cars that run on new forms of energy, (2) further developing and disseminating clean-energy automobiles and fuel-efficient automobiles, and (3) promoting associated technologies, such as lighter car bodies.

Although we may look forward to steady improvements in existing technologies – such as gas–electric vehicles, plug-in hybrid vehicles, and clean diesel vehicles – as well as to the development of various core technologies for improving fuel economy, nonetheless the severity of the global energy problem is expected to worsen around 2020 due to limited petroleum resources, and this may further promote the growth of alternative (non-oil) energy sources.

The two core technologies that hold the key to solving our energy problems are (1) battery technologies, such as high-capacity, ultrahigh-performance secondary batteries, and (2) new fuel infrastructure technologies, such as hydrogen storage. While the energy efficiency of internal combustion engines has essentially reached its theoretical limit and has little room for improvement, there is still room to improve the energy efficiency of electric devices. In particular, there are high expectations for technologies to use and accumulate electrical energy.

In parallel with the diversification of energy sources, life cycle assessments (LCA) – which address not only the environmental damage caused by vehicles in operation, but also the impact of vehicle manufacturing and that of fuel production, transport, and conversion – will become more important than in the past. We will have to consider the finite availability, safety, and environmental impact of the resources employed in new technologies, such as rare metals.

A key challenge for automobile technology policy will be the establishment of middle- and long-term energy strategies and initiatives, including improvements in large-scale fuel-provision infrastructures beyond those currently in place for gasoline, diesel oil, and LPG. As there are numerous technical options for how to proceed here, quantitative policy assessments will be critical in determining the optimal course of action.

Technologies to combat air pollution and reduce automobile noise have already reached a certain state of advancement, and are expected to continue to develop steadily under existing technology forecasts. For air pollution, the most pressing problem at present is to develop technologies applicable to the new 'post-long-term' regulatory framework. Other mid- and long-term technological challenges include measures to introduce alternative (non-oil) fuels and further to reduce PM emissions. The shift toward electric-powered vehicles will bring dramatic improvements in air pollution, as such vehicles have zero emissions at travel-time.

b) Technological and policy trends for batteries and other energy-storage devices.

As discussed above, the shift toward electric-powered vehicles is expected to bring about significant mid- and long-term alleviation of environmental and energy-related problems, but rapid technology innovations will be needed before such vehicles can attain more widespread acceptance. Batteries are the one component common to all currently proposed electric vehicle technologies, and improvements in battery function and cost are critical for the advancement of electric vehicles. Here, we will survey technological and policy trends in the development of modern high-performance batteries.

The batteries currently used to start the engine and power the accessories (such as the headlights) in ordinary vehicles are lead storage batteries. On the other hand, hybrid vehicles, which require higher energy output and storage capacity, instead use nickel-metal-hydride (NiMH) batteries. Lithium-ion (Li-ion) batteries, which boast high energy output and capacity per unit weight, have also recently begun to be used in new electric vehicles. However, neither of these types of battery is sufficiently safe, durable, and high-performing to form the basis for a significant expansion of electric vehicles, and further technological development is thus required.

Figure 2.4, based on a paper from the Japanese Ministry of Economy, Trade and Industry (METI, 2006), outlines some basic concepts in battery performance. Storage capacity per unit weight (weight-energy density) is plotted on the horizontal axis, while the vertical axis plots energy output per unit weight (weight-output density). Points in the upper right corner of the plot indicate high battery performance. NiMH batteries lie in the lower left corner, indicating that gas–electric and plug-in hybrid vehicles require batteries with both higher energy density and higher output density. The target for 2015 is for plug-in hybrids and commuter gas–electric hybrids to be able to travel 40 km and 100 km, respectively, on a single battery charge. The target suggests that realizing such travel distances will require improvements in weight-energy density to around 150 per cent of current levels. However, as batteries will also be required to have higher capacities, the target further requires costs to fall to one-seventh of current levels. Attaining these targets will require sophisticated technology development, including a comprehensive review of battery structure and materials.

On the other hand, dramatic technology innovations will also be required to achieve widespread use of high-performance, full-range electric vehicles. The target for full-range electric vehicles is to travel a distance of 480 km on a single battery charge while boasting the same performance levels as current gasoline-powered vehicles. To achieve these targets, battery energy densities must increase seven-fold, while costs must be reduced by a factor of 40. The performance limit of lithium-ion batteries is thought to be an energy density of around 250 Wh/kg; realizing this limit will require discovering new substances to act as carriers of electricity and other fundamental scientific breakthroughs.

The plug-in hybrid vehicles currently used by a limited number of early adopters boast travel distances of 23.4 km when running on electricity alone. Clearly a yawning chasm lies between today's performance and future targets, and technological progress will depend on steady research and development efforts.

When such drastic technological innovation is required, private companies alone cannot make all necessary investments, and government support is indispensable. As innovations in battery technology are in the nation's interest – empowering industry and reinforcing global competitiveness – public support for research and development is provided not only in Japan, Europe, and the US, but also in other countries such as China and South Korea. In Japan, R&D support for battery technology is provided on a project-by-project basis, with a consortium of industrial, government, and academic participants required to achieve concrete objectives by predetermined deadlines. The same structure is used for fuel cell research and development.

On the other hand, regulatory policies, such as legislation mandating the installation of new technologies or subsidies to promote their spread, cannot by themselves achieve technological innovation targets. For example, California's ZEV (zero-emission vehicles) regulations essentially mandated the introduction of electric and fuel-cell vehicles, but the technological advancements that would be needed

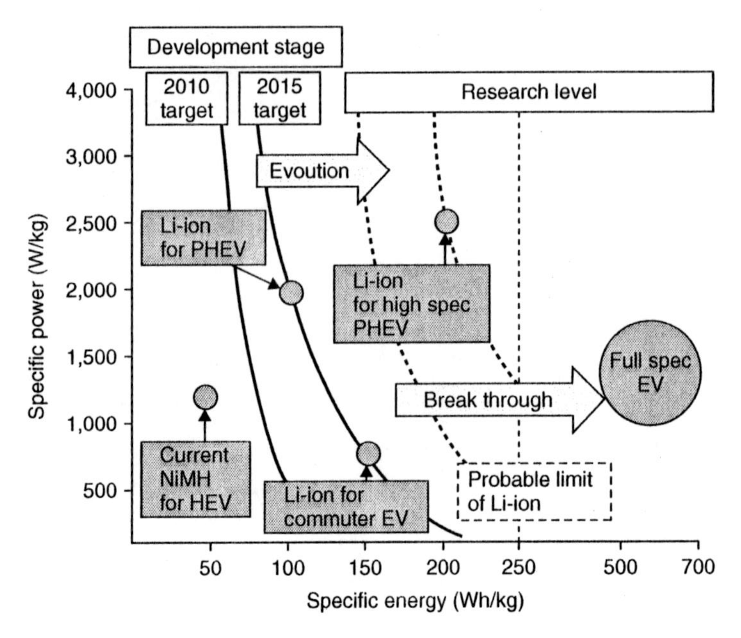

Figure 2.4 Current status of and future targets for battery performance
Source: Prepared by the authors based on document by METI (2006).

to achieve such a goal have not been made. The failure can be attributed to overly optimistic predictions of improved performance and reduced cost of core technologies. Such policies to promote the spread of technology are inefficient unless the technologies are marketable on their own, and must be adopted only when we have realistic prospects for technological innovation and accurate assessments of market viability.

2.3.2 Safety-related ITSs

a) Trends in technology development.

Traffic safety technology will remain the most important of all automotive technologies in the future. While the number of traffic fatalities in Japan is enjoying a decreasing trend thanks to various safety measures – the number of fatalities in 2008 was 5,115 – the number of injuries remains high, in excess of 945,504.[4] When we consider that damage suffered in traffic accidents is more severe for senior citizens than for younger citizens, and in view of Japan's societal trend toward an older society with fewer children, measures to reduce the number of victims of traffic accidents will clearly be required.

While traffic safety technologies in the past were developed with a primary focus on passive safety measures such as collision safety improvement, the focus in recent years has shifted to the development of active, preventative safety technologies aimed at avoiding traffic accidents in the first place. Among the core technologies underlying active safety measures, environmental recognition technologies play a particularly important role. Image processing technology is especially critical as a core technology for collision avoidance and for recognition of the vehicle's operating environment, and car companies are competing with one another to develop such technologies. In the future we can expect further technological developments in this area, including improved sensor technologies and the development of high-level information and communication technologies to anticipate drivers' intentions.

Here, we will review trends in the development and roll-out of ITS technologies in the traffic safety sector, which will play a leading role in future active safety technologies. Although Japan's communication infrastructure is well developed, the nation has not yet reached the point at which this infrastructure can be utilized to full advantage by moving cars. Active safety technologies which have already gone into practical use include the autonomous detection-type driving support systems, which are self-contained systems installed within an individual car. This system includes three distinct vehicle safety subsystems: (1) an automated following-distance control system, which enables vehicles automatically to maintain proper inter-vehicle distance by detecting the car ahead (within the same lane) and by applying the brake accordingly; (2) a rear-end collision avoidance system, which applies the brakes to avoid collisions with the car ahead; and (3) an automatic lane-maintenance assistance system, which detects white lines on the road and alerts the driver when the vehicle risks deviating from its lane. These systems provide assistance in negotiating situations that can be visually recognized by drivers. Other safety-related ITS technologies drawing recent attention include the *roadside-information-based* driving support systems (hereafter referred as to car-to-infrastructure systems), based on road-to-vehicle communication, and the *inter-vehicle-communication-based* driving support systems (hereafter referred as to car-to-car systems), based on inter-vehicle and vehicle-to-road-to-vehicle communication. The car-to-infrastructure systems use roadside detectors to identify cars or pedestrians outside the driver's field of vision and use roadside transmitters to notify drivers of these obstacles. The car-to-car systems avoid accidents by allowing exchange of information between cars at risk of collision or by sending information received from other vehicles to cars at risk of accident. These systems differ from the autonomous detection-type driving support systems in that they assist drivers in negotiating situations that cannot be easily recognized or cannot be visually recognized at all. The autonomous detection-type driving support systems will be effective in preventing frontal or rear-end collisions, but have limited impact in reducing accidents at intersections or crashes with pedestrians or bicycles. The car-to-infrastructure systems and car-to-car systems are more effective in preventing such accidents.

In January 2006, Japan's Cabinet Office announced a 'Strategy for IT Reform' intended to form the basis of Japanese information technology (IT) policy in the future. The strategy identifies 'the realization of the world's safest road traffic system' as one of its 15 major themes, and sets objectives such as putting IT-based traffic safety services into practical use by 2010. Development and experimental testing of the car-to-infrastructure systems and the car-to-car systems in Japan are currently conducted through the collaborative efforts of three entities: the Ministry of Land, Infrastructure, Transport and Tourism, the National Police Agency, and car-makers.

b) Future policy directions.

In comparison with the car-to-car systems, the car-to-infrastructure systems have the advantage that they can begin immediately to have an effect without waiting for installation of on-board communication devices to spread throughout society, but suffer from the drawback that they cannot be used on roads on which the roadside devices are not installed. On the other hand, the car-to-car systems (excluding their vehicle-to-road-to-vehicle communication features) have the advantage of being usable irrespective of location, as long as drivers are equipped with the on-board communication device, as this does not need roadside devices to function. However, a disadvantage of this system is that it has little effect unless many cars are equipped with the on-board devices.

There is thus at present a movement toward launching the car-to-infrastructure systems, with information to be transmitted through an optical-beacon-compliant car navigation system, making use of some 50,000 optical beacons already installed throughout Japan for the VICS system. However, the optical-beacon version of the system suffers from limited transmission rate and reduced maximum communication distance, allowing information to be sent only through simple images or sound effects. An alternative, considered until recently, was to introduce a system known as the Dedicated Short Range Communication (DSRC), an ITS communication scheme using the 5.8 GHz broadcast band. This method would allow vehicles turning right at an intersection to see an image of any cars that might be approaching the intersection and planning to pass through. The scenario anticipated by many IT experts was thus that the car-to-car systems would be gradually introduced, with DSRC-compliant on-board devices expected to spread throughout the consumer base.

However, with the shift to digital schemes for terrestrial television broadcasting, it was decided that the 720 MHz band, which had been used for analog TV broadcasts, would instead be allocated to ITSs, beginning on July 25, 2012. Indeed, because Ultra High Frequency (UHF) band can cover a wider geographical area and has a longer diffraction distance than the 5.8 GHz band used by DSRC, the UHF band is more appropriate than DSRC for the car-to-car system, which is designed to communicate with automobiles which may be hidden behind bridges, buildings, or other structures and which are distributed over a large geographical area. The Ministry of Internal Affairs and Communications (2009) recommends that both of these car-to-car and car-to-infrastructure systems should be accessible from one on-board device for the benefit of end-users (see Chapter 7).

In any case, a strong broadband communications infrastructure, which enables vehicle users seamlessly to obtain and process information via road-to-vehicle and vehicle-to-road-to-vehicle communication, and to receive information from various unspecified information sources without special operations or on-board information devices, will clearly be required in the future.

When it comes to drafting technological policies for this sector, beyond the key questions of wireless communication bands, policies addressing institutional and economic matters will be critical. On the institutional side, legal measures addressing responsibility at the time of an accident must be reconsidered. Indeed, Japan's Automobile Liability Security Law is premised on the underlying assumption that drivers control automobiles; in a world of safety-related ITS technologies, this notion falters when we consider drive assistance systems and breaks down entirely in the case of automated drive intervention. The penetration of these systems throughout society thus cannot advance until an appropriate legal framework, addressing in particular the interrelationship with the Product Liability Law, has been constructed.[6] On the economic side, for safety-related ITSs to attain widespread use throughout society will require carefully constructed promotional policies that take into account the unique economic features of these systems, namely, their network externality. This point will be discussed in detail in Chapters 5 and 7.

2.4 Conclusions

Automobiles fall within the most expensive category of general durable consumer goods, and automotive technologies have the features that (1) they involve the aggregation of numerous core technologies, (2) they take a long time to develop, (3) they require large-scale investment, (4) products take about 10 years to achieve significant market penetration, and (5) long-term durability and high safety standards are strongly desired attributes in the marketplace. In view of these facts, it takes longer for the fruits of automotive technology innovations to penetrate the market, and longer lead times are thus required for automotive technology development than is the case for other products, such as consumer electronics or household appliances. Mid- and long-term technological expectations are thus crucial in determining appropriate directions for investment. This is a key point to understand when crafting future automotive technology policies. Considering trends in the mid- and long-term development of automobile technologies, extending to around the year 2030, the priorities for technological innovation will be to protect the environment (particularly the global environment) and improve traffic safety, and addressing these challenges will spur the overall technological advancement of the industry.

Future automotive technology policies must be formulated in such a way as to promote technological innovation. The automotive technologies required to address environmental and safety problems will encompass not only core automobile technologies but also various other technologies relating to the environment in which vehicles are used. In addition, it will be increasingly important to establish social systems, such as traffic and recycling systems, and effective use of information and communication technologies will grow ever more critical. The challenge for the future will be to choose the most appropriate policy from the ever-expanding set of possible policy options, which are more numerous today than ever before.

As for the *form* taken by automotive technology policies, although it will certainly continue to be necessary to institute regulatory policies for the guarantee of safety and reliability of future automobiles, we should also consider expanding the use of economic measures, thus further expanding the already numerous set of available policy options. In drafting economic measures and building consensus among the parties affected by new policies, policy assessments based on quantitative analysis methods, such as cost–effect analysis, will be indispensable.

Examples of such quantitative assessments of automotive technology policy have been rare in Japan, and in the past no quantitative assessments have been made prior to the introduction of regulatory policies (Institute of Administrative Management, 2004). On the other hand, other countries have conducted numerous attempts at quantitative assessment of automotive technology policies, such as the early study by Weitzman (1974), who carried out a comparative analysis of regulatory and economic methods in environmental policies. This is probably due to the fact that Regulatory Impact Analysis (RIA), which requires that the necessity of regulations be assessed and that the associated costs and benefits be compared with alternatives (OECD, 2009), was systematized from the earliest stages.[7]

In Japan, although we can cite the pioneering experimental analysis of Hamamoto (1997) on the relationship between environmental regulations and R&D expenditures, the number of such studies has been limited. However, as mentioned at the end of Section 2.3, the *public comment* system mandated by the revised rules for administrative procedure has enabled researchers to obtain detailed information on the reasoning behind regulatory policies, and there have recently begun to appear some experimental studies related to regulatory policies, including cost–effect and cost–benefit analyses (Fujiwara, 2006; Kainou, 2007; Kii *et al.*, 2007; and see also Chapter 3 of this book). We hope that quantitative assessments and experimental analyses of automotive technology policies will become more common in the future.

With regard to internal combustion engines and their attendant problems, we anticipate no significant problems in continuing to entrust technology development to those who have traditionally shouldered this responsibility, namely, automotive engineers. However, other increasingly important challenges, including new technologies that depend heavily on information and communication technology infrastructures, as well as the design of societal systems such as recycling systems and the construction of environmentally conscious urban structures, cannot be solved through technological innovation alone. To maximize social benefit we must thus strive to balance the involvement of *players with a stake in technology development* and *players with a stake in the results of technology development* in the process of adopting new technologies, and future automotive technology policies must be designed with this objective in mind.

Notes

1. See Ralph Nader (the American social activist and lawyer), *Unsafe at Any Speed: The Designed-In Dangers of the American Automobile* (1965) and other sources.
2. Before the establishment of the standard, the Ministry of Transport formulated a 'Long-term plan for the prevention of toxic automobile emissions' in 1964. Similarly, the automobile industry had been engaged in an examination of regulatory methods and standards through its 'Round-table conference on automobiles and the air pollution problem in Japan', also organized in 1964, and by dispatching a 'research group on automobile pollution' to the US in 1966.
3. ART Team: Isuzu Motors, Suzuki Motor, DaimlerChrysler Japan, Nissan Motor, Nissan Diesel Motor, PAG Import, Ford Japan, Fuji Heavy Industries, MAZDA Motor, Mitsubishi Motors, and Mitsubishi Fuso Truck & Bus.
   TH Team: Daihatsu Motor, Toyota Motor, Hino Motors, Honda Motor, Audi Japan, BMW, Peugeot Japan, and Volkswagen group Japan.
4. On this topic we have referred to a lecture entitled 'ITS from the legal perspective', delivered by Professor Tomonobu Yamashita of Tokyo University at the Seventh ITS Symposium, held on December 5, 2008.
5. Source: Traffic Bureau, National Police Agency.
6. After its institutionalization in the US in the 1970s, RIA was institutionalized in the UK and Canada in the 1980s and later in Australia and New Zealand in 1995. Since the second half of the 1990s it has been institutionalized in most OECD (Organization for Economic Co-operation and Development) countries.

References

Fujiwara, T. (2006) *'Syouhisha no sinsha kounyuu koudou ni taisuru jidousha green zeisei no eikyou hyouka'* (Assessment of Automobile Green Taxation System's Impacts on Behaviors of Consumers Purchasing New Cars), in Institute for Policy Sciences (ed.), *Heisei 17 nendo keizai sangyosho itaku chousa, Heisei 17 nendo kokusai enerugi siyou goorika kiban seibi jigyou houkokusho (chikyuu ondanka no hiyoutai kouka nikansuru seisakuhyouka chousa)* (2005 Research Entrusted by the Ministry of Economy, Trade and Industry, Report on International Infrastructure Development Project Concerning the Rational Use of Energy (Policy Assessment Concerning the Cost-Effectiveness of Measures Against Global Warming)), pp. 40–53 (in Japanese).

Hamamoto, M. (1997) *'Porter kasetsu wo meguru ronsou ni kansuru kousatsu to jisshou bunseki'* (Studies and Experimental Analyses of Discussions Concerning the Porter Assumption), in *Keizai Ronsou* (Kyoto University's Economic Review) 160(5–6): 102–20 (in Japanese).

Institute of Administrative Management (2004) *Kisei hyouka no hurontia – kaigai ni okeru kisei eikyou bunseki (RIA) no doukou* (Frontier of Regulatory

Assessment: Foreign Trends of Regulatory Impact Analyses (RIA)), Tokyo: Institute of Administrative Management (in Japanese).

Kainou, K. (2007) *'Toppu rannar houshiki' ni yoru shouenerugii hou jouyousha nenpikisei no hiyoubeneki bunseki to teiryo teki seisaku hyouka ni tuite* (Concerning the Cost–Benefit Analysis and the Quantitative Policy Assessment of Energy-Savings Fuel Regulations for Passenger Vehicles by Leading Runner Approach), RIETI Discussion Paper Series, 07-J-006 (in Japanese).

Kii, M., Kameoka, A., Hosoi, K. and Minato, K. (2007) *'Jouyousha nenpi kaizen gijutsu dounyuu kouka no ichi kousatsu'* (Impact Assessment of Fuel-Efficient Technologies for Passenger Vehicles), *Energy and Resources*, 28(3): 168–74 (in Japanese).

Ministry of Economy, Trade, and Industry (METI) (2006) *Jisedai jidoushayou denchi no shourai ni muketa teigen* (Recommendation for the Future Battery for Next Generation Vehicles), *Sinsedai jidousha no kiso to naru jisedai denchi gijutsu ni kansuru kenkyuukai* (Research Committee on Next Generation Battery Technologies for the Basis of New Generation Automobiles), Tokyo: METI (in Japanese).

Ministry of Environment (MOE) Website, available at http://www.env.go.jp/en/wpaper/ (accessed November 26, 2010).

Ministry of Internal Affairs and Communications (MIC) (2009) *ITS musen shisutemu no koudoka ni kansuru kenkyu kai houkoku sho* (Report of the Research Group on the Evolution of High-Level Wireless ITS Technology), available at http://www.soumu.go.jp/menu_news/s-news/14422.html (accessed November 26, 2010) (in Japanese).

Ministry of Land, Infrastructure, Transport and Tourism Website, available at http://www.mlit.go.jp/road/ITS/ (accessed November 26, 2010).

Nader, R. (1965) *Unsafe at Any Speed: The Designed-In Dangers of the American Automobile*, New York: Grossman Publishers.

OECD (2009) *Regulatory Impact Analysis: A Tool for Policy Coherence*, Paris: OECD Publishing.

Society of Automotive Engineers of Japan, Inc. (2007) *Nihon no jidousha gijutsu 240 sen* (Japanese Automotive Technology Selections 240), available at http://www.jsae.or.jp/autotech/ (accessed November 26, 2010) (in Japanese).

The Institute of Energy Economics, Japan (2010) *EDMC Handbook of Energy & Economic Statistics in Japan 2010*, Tokyo: The Institute of Energy Economics, Japan.

Weitzman, M. L. (1974) 'Prices vs. Quantities', *Review of Economics Studies*, 41(4): 477–91.

3

Costs and Benefits of Fuel Economy Regulations: A Comparative Analysis of Automotive Fuel Economy Standards and the Corresponding Benefits for Consumers

Masanobu Kii and Hiroaki Miyoshi

3.1 Introduction

Improving automobile fuel economy is one of the most promising approaches to saving energy and reducing CO_2 emissions in the transport sector. Automotive fuel economy regulations have been established in many countries as policy measures designed to bring about such an improvement, but the specific implementation of these regulations takes different forms in different countries and regions: the Corporate Average Fuel Economy (CAFE) standards in the US, voluntary agreements in the European Union, average fuel economy standards by inertial weight (IW) class in Japan, and minimum requirements for fuel economy by weight class in China. A clear understanding of the advantages and disadvantages of each approach, together with an analysis of effective market conditions, will be of use to countries seeking to establish or update their own regulations.

Consumer acceptance of such regulations is another important issue. New technologies to improve automobile fuel economy tend to raise vehicle prices, but the greater fuel economy tends to reduce fuel costs. The economic benefit of fuel economy improvements for consumers thus depends sensitively on the relationship between increased vehicle prices and reduced fuel costs during the vehicle's lifetime. This, in turn, means that well-crafted regulations can simultaneously benefit consumers and reduce CO_2 emissions.

We begin this chapter by reviewing the major fuel economy regulatory structures in place around the world, and by comparing the different effects of different approaches, taking as specific examples the cases of CAFE standards in the US and IW class regulations in Japan. As an evaluation metric, we use estimated vehicle price, which is one of the most important signals from manufacturers to consumers. Next, we evaluate the effect of regulated fuel economy standard levels on vehicle price. We then analyze the benefits to consumers of fuel economy regulations, considering the specific case of weight class-based fuel economy regulations in Japan. Finally, based on the results of this analysis, we discuss issues relevant to future regulations.

3.2 Automotive fuel economy regulations

Standards for fuel economy and/or CO_2 emissions for new vehicles have been proposed and established in several countries and regions around the world. However, the regulatory approaches are quite different in different regions, especially among the three largest automobile markets – the US, the European Union, and Japan. Although the regulatory approaches taken often depend on historical or political factors, when introducing new standards or revising current standards it is nonetheless beneficial to understand the merits and demerits of each approach. In this chapter we describe the regulatory approaches taken in the major markets, based on An *et al.* (2007), An and Sauer (2004), and Plotkin *et al.* (2002), as well as on other surveys in the literature.

The US CAFE program sets standards for the sales-weighted fuel economy of light-duty passenger vehicles. The program distinguishes passenger cars from light trucks and applies different standards to each. Until recently, the standards had remained unchanged since 1985 for passenger cars and since 1997 for light trucks. It was decided to raise the standard for light trucks in miles per gallon after 2005, and a vehicle-footprint-based standard was introduced starting in 2008. This approach takes vehicle size into account to arrive at a new standard designed to alleviate the influence of unexpected changes in market preferences on sales-averaged fuel economy.

The US State of California had taken a similar approach, namely the sold-fleet-average standard, but applied the standard to CO_2 emissions from the tailpipe. An emission standard measured in grams of CO_2 per mile was introduced in 2009, with plans to tighten the standard every year until 2016. This regulation had virtually the same effect as the CAFE, but with standard levels significantly more stringent than the federal regulations. This regulation has been stalled due to lawsuits by auto-makers and by the lack of an implementation waiver from the US Environmental Protection Agency (EPA).

In response to President Obama's 2009 call to reduce greenhouse gas (GHG) emissions, the US EPA and DOT (Department of Transportation) finalized joint regulations for GHG emissions and fuel economy in April 2010. These regulations reflect an agreement between the US federal government, the State of California, and auto manufacturers regarding a unified national program to regulate automobile GHG emissions and fuel economy (International Council on Clean Transportation, 2010). Under this agreement, EPA will regulate GHG emissions, while DOT will simultaneously establish corresponding CAFE standards. The regulations are formulated as piecewise-linear relations between vehicle size (footprint) and GHG emission rate, with smaller vehicles subject to more stringent standards. Whereas previous CAFE standards were designed to regulate only the weighted average fuel economy of all vehicles sold, these new regulations are designed to reflect the sizes of vehicles sold. It is the same approach as was introduced for light trucks in 2008. In the new CAFE, the vehicle footprint-based standard is applied to passenger cars as well and the level of standard is updated. Thus, in contrast to previous CAFE approaches, these newer regulations do not allow manufacturers simply to promote sales of smaller vehicles in order to meet CAFE standards.

The European Commission has adopted voluntary agreements (VAs) with automobile industry groups, including ACEA (Association des Constructeurs Européens d'Automobiles), KAMA (the Korean Automobile Manufacturers Association) and JAMA (the Japanese Automobile Manufacturers Association). These agreements require average CO_2 emissions from new vehicles to be 140g CO_2/km or less by 2008. This target was collective in nature and applied to the average of all vehicles sold by the member companies of each group (ACEA, KAMA, and JAMA). The original idea of this approach was that peer pressure within each group would force auto-makers to improve fuel economy at least to the levels attained by other member companies, particularly in cases where significant differences in fuel economy existed between companies in the same group, as in the case of Renault and BMW. The expectation was that the group-average target structure would distribute differences in fuel economy flexibly and would achieve fuel economy goals at lower cost. On the other hand, the collective target renders the responsibility of each individual company somewhat ambiguous. In part because of increased sales of larger automobiles such as SUVs, the goals for 2008 were not met. This failure prompted the establishment, in April 2009, of legislation regulating CO_2 emissions from passenger cars – a mandatory scheme with financial penalties levied on excess emissions (European Commission, 2010). The emission targets reflect vehicle weight – heavier vehicles are allowed higher emissions – while the region-wide average to be achieved is set at 130g CO_2/km, where the base weight for the calculation of vehicle-specific emissions seems to have been adjusted to meet 2016 targets.

The Canadian and Australian approaches, in contrast, are voluntary, not mandatory. However, in practice, the Canadian approach has compelling force. The Canadian Parliament passed legislation for a mandatory program to reduce vehicle fuel consumption in 1982, although the legislation never went into effect because the vehicle industry agreed to comply voluntarily with the standards of the act. The standard applies to corporate average fuel consumption; that is, each company has responsibility and the government still has the option of legislation. The impact of this program thus seems, in essence, to be quite similar to that of the US CAFE program. A further similarity with the US CAFE program is that the target fuel consumption level has remained unchanged for passenger cars since 1985 and for light trucks since 1995.

In the Australian approach, the government and the Federal Chamber of Automotive Industries (FCAIs) have agreed upon a voluntary target. The two bodies had also agreed upon such voluntary targets on two previous occasions, but in both cases the target was not achieved within the intended interval. Notwithstanding this apparent failure, the voluntary approach is believed to have contributed to the reduction of fleet-wide fuel consumption. The current fleet-average target for passenger cars is to reduce fuel consumption by 18 per cent between 2002 and 2010. This is also a voluntary scheme, and no specific enforcement or non-compliance penalties have been proposed in the case of failure.

In Japan, based on the revised 'Law Concerning the Rational Use of Energy', fuel economy targets for passenger cars by inertial weight (IW) class were established for the year 2010. Recently, rules concerning targets for the year 2015 were enacted in 2007, on the basis of the most fuel efficient vehicles in the market in the benchmark year based on the top-runner program. The targets are determined for each IW class and are to be achieved in sales-average by individual companies. The number of categories will be almost doubled, from nine in 2010 to 16 in

2015. In contrast to the US CAFE and European voluntary agreements, these standards are based on vehicle weight, which is correlated with fuel economy. This approach has the virtue of equalizing technological challenges for each company by alleviating the influence of differences in sold-vehicle weights among companies. On the other hand, targets by weight class provide little incentive to promote smaller vehicles or reduced weight, because more stringent targets will be applied to lighter classes. In addition, unlike the US and EU approaches, the Japanese approach does not guarantee any improvement in average fuel economy when a company changes its distribution of sales by weight class.

China also has standards based on vehicle test weight, but – in contrast to other countries' standards regulating sales-average fuel economy – the Chinese approach regulates maximum limits for fuel consumption by new vehicles. As these standards are designed to put more stringent limits on heavier vehicle classes than lighter ones, we may think of this approach as an incentive to manufacturers to produce lighter vehicles. On the other hand, in contrast to sales-average regulations, the Chinese approach provides no incentive to make further improvements to the fuel economy of vehicles already within the regulatory limits.

Both Taiwan and South Korea have standards based on the displacement volume of automotive engines. However, the two countries take different approaches: Taiwan – like mainland China – has regulated maximum fuel consumption levels since 2004, whereas South Korea uses average fuel economy standards. Taiwan sets standards for seven categories by engine size, and vehicles exceeding these limits are not approved for sale within the country. South Korea replaced voluntary standards with mandatory standards in 2004, and the number of categories was reduced from eight in the voluntary system to two at present. The current standards were introduced in 2006 for domestic vehicles, and in 2009 for imports. The vehicle category is bounded at 1500 cc of engine displacement. The average fuel economy of new passenger cars in 2008 is 11.47 km/l, improved from 11.04 km/l in 2007. According to the Ministry of Knowledge Economy in South Korea, this improvement is mainly due to increasing sales of lightweight or small cars, as well as to advances in engine and parts technologies.

As this survey demonstrates, multiple approaches and methods have been established around the world to regulate vehicle fuel economy, based on policy targets and on the maturity and structure of the automobile industry, as well as on government administration, market preferences and trends, and technological progress. In addition to differences between mandatory and voluntary schemes, other policy measures, such as the introduction of taxes or subsidies and labeling schemes, may be considered to make standards effective in the real market. Although not mentioned above, technical issues, such as the test cycle used to measure fuel consumption, are also important in establishing standards. Although there are many issues to consider in detail, here we will focus on the effects of categorization of vehicles for standards, as this is one of the most significant differences among the various approaches and has significant impact on the cost-effectiveness of the measures adopted.

In Sections 3.3 to 3.7, we compare the CAFE standards of previous years – which were independent of vehicle weight – against weight class standards, with an eye toward effects on vehicle cost, the most important signal from manufacturer to consumer. As discussed above, the US CAFE standards have recently been modified to consider the sizes of vehicles sold. We can interpret this to mean that the new CAFE regulations incorporate some features of Japan's IW class regulations – particularly those features that seek to alleviate the impact of shifting market preferences toward larger vehicles on compliance with the standards. In these sections, we would like to clarify the essential features of CAFE and IW class standards, and thus we take past CAFE standards as a basis for comparison. Thus, throughout this chapter, the term CAFE refers to *past* implementations of CAFE standards. In Section 3.3, we describe the effects of each approach on manufacturers' use of technologies. Then we compare prices for vehicles meeting the targets under these two approaches over a range of target levels.

3.3 Technological strategies for meeting CAFE and weight class standards

These two approaches, CAFE and IW class regulations, are thought to have different effects on the technology strategies chosen by automakers, due in particular to two factors: changes in the distribution of sales by vehicle weight, and weight reduction technologies. The two approaches have significantly different impacts on the ability of automakers to achieve standards. We thus compare the effects of the two approaches with particular focus on changes in sales distribution by weight class and on weight reduction technologies. We then summarize the effects of the two approaches on the strategies chosen by automobile producers to meet the standards.

A change in sales distribution by vehicle weight will have a larger effect on compliance with CAFE standards than on fuel economy by IW class. If light-weight vehicles represent a larger percentage of total sales, then the CAFE standards are easier to meet, because lighter-weight vehicles are more fuel efficient. In contrast, the larger the percentage of total sales occupied by heavy-weight vehicles, the more difficulty the manufacturer will have in meeting the standard. It is possible to apply marketing or pricing strategies to increase the percentage of total sales commanded by light-weight vehicles, although the sales distribution ultimately reflects consumer preferences, which auto-makers are not always able to control. Plotkin *et al.* (2002) pointed out that controlling vehicle prices in order to lead the market is a high-cost measure for auto-makers. Under the CAFE standard, auto-makers may have to consider the risk of non-compliance with the standards due to changes in the sales distribution.

Figure 3.1 shows the CAFE standard for passenger cars and the observed sales-average fuel economy of selected companies in the US. The fuel economy of each company fluctuates from year to year; some companies exhibit occasional dips below the level of the standard. The

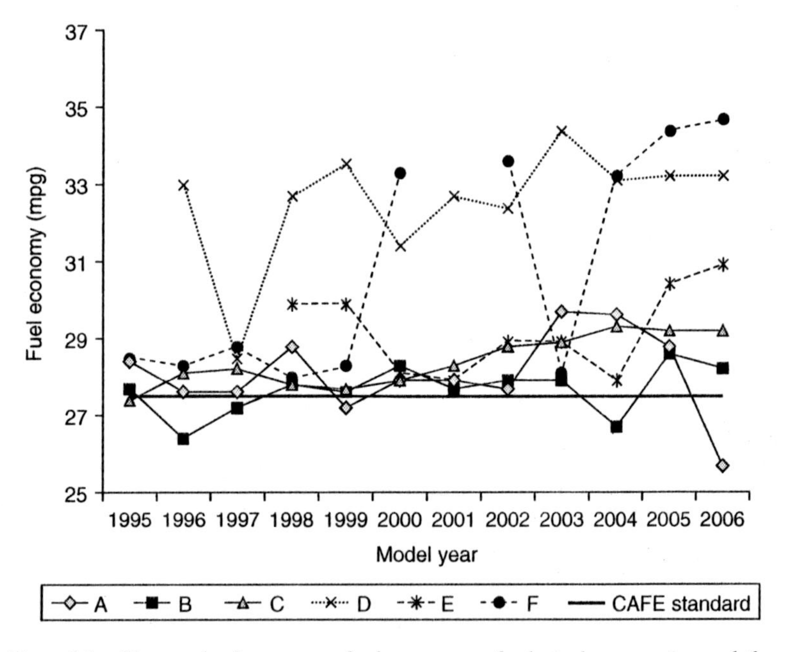

Figure 3.1 Observed sales-average fuel economy of selected companies and the CAFE standard in the US

Source: Based on National Highway Traffic Safety Administration (2006).

program allows credit transfers within the previous and following three years, and companies do not necessarily have to pay immediate penalties for failure to meet the standard. For some companies, fuel economy levels fluctuate more than 15 per cent at certain points. It seems likely that this behavior is caused not by automobile technologies but by changes in sales distribution. Some companies' average fuel economies far exceed the standard, which probably reflects consumer preference for vehicles with better fuel economy, as well as the risk of consumers abandoning companies that fail to comply with the standard.

Weight reduction technologies are limited under IW class regulations as compared with the CAFE (Plotkin *et al.*, 2002). Figure 3.2 shows Japanese fuel economy standards by IW class for 2015. The plot indicates more stringent standards for lighter classes. For instance, if weight reduction technologies move a vehicle into a lighter weight class, the vehicle then has to meet a more stringent standard. Weight reduction technologies for improvements in fuel economy would thus be restricted to a limited number of vehicles. The Japanese standards for 2015 have almost twice as many weight classes as do the 2010 standards, and the differences in standards between adjacent classes are reduced. Nonetheless, weight class regulations will still place constraints on the utilization of weight reduction technologies. In contrast, CAFE standards do nothing to discourage auto-makers from employing

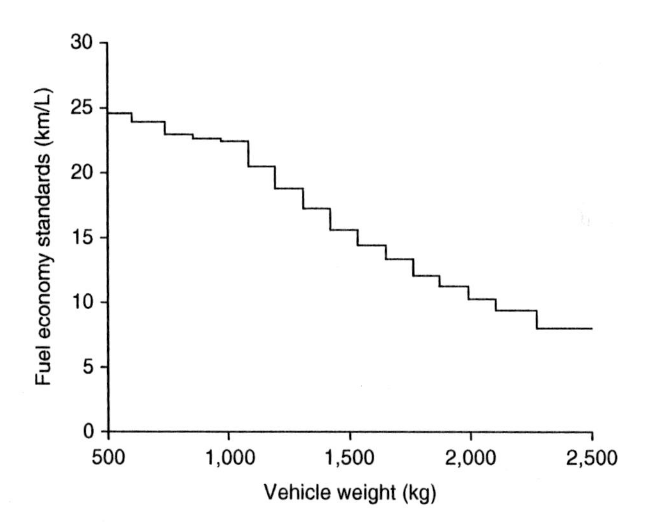

Figure 3.2 Japanese fuel economy standards for 2015
Source: Ministry of Economy, Trade and Inductry (2007).

weight reduction technologies, because the standards are independent of vehicle weight.

The two approaches to fuel economy regulation, CAFE and IW class standards, are expected to have different effects on the technology strategies chosen by automobile producers. The new CAFE standards, which take vehicle footprint into consideration, may alleviate the tendency of market fluctuations to cause compliance failures, but may also limit the utilization of weight reduction technologies. In this sense, the new CAFE standards have incorporated some features of vehicle-class-based standards. Thus, for clarity, we should compare the advantages and disadvantages of the *original* CAFE standards with those of weight class standards. In this paper, automobile companies are assumed a) to control the risks of non-compliance caused by changes in sales distribution under the CAFE program, and b) to use no weight reduction technologies to improve fuel economy under the IW class regulations. In this study, due to data limitations, we assume three weight classes: mini, small, and ordinary.

3.4 Methods for assessing the impact of regulation on vehicle price

In this section, we estimate the impact of regulations on vehicle price. First, we summarize our assumptions regarding the anticipated impact of each regulatory approach on auto-makers' technological strategies. Second, we formulate methods for implementing technology strategies under each approach. Third, we quantify the cost of improvements in fuel economy. Finally, we analyze the impact on vehicle cost and the corresponding price sensitivity. Here the cost is measured by Retail Price Equivalents (RPE), and we do not distinguish the value of cost and price in this chapter.

3.4.1 Assumptions

We assume that auto-makers will utilize fuel efficiency technologies to minimize vehicle price under the fuel economy constraints imposed by regulations. We assume no technological differences among auto-makers. When auto-makers are able to utilize weight reduction technologies, we assume that the cost of improvements in fuel economy is lower than in cases where such technologies are not available, because weight reduction technologies increase the number of options for improvement. We assume that the sales distribution by weight class fluctuates stochastically. We assume that auto-makers under the CAFE regulations utilize fuel efficiency technologies to meet the standard with a certain probability. These assumptions are summarized in Table 3.1.

Table 3.1 Advantages and disadvantages assumed for each regulatory approach

	Weight reduction technology	**Change of sales distribution by weight class**
Weight class standards	Unavailable	No risk of non-compliance
CAFE standards	Available	Risk of non-compliance

3.4.2 Formulation of auto-makers' technological strategies

a) Weight class regulation.

The weight class is denoted by g ($\in G$: all classes), the base fuel consumption (L/km) by e_{0g}, the target improvement rate by α_g, and the cost function without weight reduction technologies by f_{Ng} (α_g). With the target fuel consumption standard denoted by e_g*, α_g is derived from the following formula:

$$\min f_{Ng}\left(\alpha_g\right),$$

$$\text{s.t. } e_g{}^* \geq \left(1-\alpha_g\right)\cdot e_{0g}. \tag{3.1}$$

Here, when $f_{Ng}(\alpha_g)$ is assumed to be a monotonically increasing function, the improvement rate $\alpha_g{}^*$ that minimizes the cost is given by the equation:

$$\alpha_g{}^* = 1 - e_g{}^*/e_{0g}. \tag{3.2}$$

The additional cost of the vehicle is denoted as $f_{Ng}(\alpha_g{}^*)$. Under weight class regulations, when the standard for each weight class is fixed, auto-makers can meet the target only by focusing on technology development.

b) CAFE regulation.

The cost function with weight reduction technology is denoted by $f_{Wg}(\alpha_g)$. The sales distribution by weight class is expressed as a vector $\mathbf{r}=\{r_g \mid g\in G\}$ with probability density $p(\mathbf{r})$. With the CAFE standard denoted by e^*, the improvement rate by weight class α_g must satisfy the following inequality:

$$e^* \geq \sum_{g\in G} r_g \cdot \left(1-\alpha_g\right)\cdot e_{0g}. \tag{3.3}$$

Under given e^*, e_{0g}, and α_g, the set of weight class fractions $\mathbf{r}$ that satisfy Eq. (3.3) is denoted by $\mathrm{R}(\boldsymbol{\alpha},\mathbf{e}_0,e^*)$, where $\boldsymbol{\alpha}=\{\alpha_g|g\in G\}$ and $\mathbf{e}_0=\{e_{0g}|g\in G\}$. If auto-makers are assumed to choose their fuel economy improvement rate to achieve the CAFE standard with probability at least P^* while minimizing vehicle price, then the fuel economy improvement rate may be obtained by solving the following formula:

$$\min \sum_{g\in G} r_g \cdot f_{Wg}\left(\alpha_g\right),$$

$$\text{s.t. } P^* \leq \int_{\mathrm{R}(\boldsymbol{\alpha},\mathbf{e_0},e^*)} p\left(\mathbf{r}\right) d\mathbf{r}. \tag{3.4}$$

In this formula, lower rates of fuel economy improvement reduce the average cost but also decrease the probability of meeting the standard. Auto-makers are assumed to control only the fuel economy improvement rate in the above equations, while the sales distribution is considered to be an external factor, even though it affects the cost and the probability of meeting the standard. In this study we assume that auto-makers will control risks such that they will fail to meet the standard with probability at most $1-P^*$.

In the actual CAFE program, fines are levied depending on the degree of non-compliance with the standard. It would thus be possible to define an expected cost of non-compliance based on the probability of changes in the sales distribution and the fine system. However, non-compliance with the standard has other negative consequences which are more difficult to quantify, such as damage to the company's image. In actual corporate management, we would expect such risks to be taken into account, but in this study we give P^* a priori for the sake of simplicity.

3.5 Estimates of cost functions for fuel efficiency improvements

The above formulae refer to cost functions for fuel efficiency improvements with and without weight reduction technologies. Cost estimates for future technologies require various assumptions and contain significant uncertainty. The Transport Research Board and the National Research Council (2002) pointed out that the fuel economy level derived from a cost–benefit analysis should not be the exclusive goal of the regulation, but it is one of the tangible results, reflecting the assumptions of technological availability, economic feasibility, consumer behavior, and various other parameters.

Future estimates by model analyses contain essentially the same problems. It should be noted that the cost functions estimated in this study would also change if the conditions assumed were altered.

First, in this section, we compile results on fuel economy improvements and the cost of selected fuel efficiency technologies from the literature. We distinguish turbocharger technologies and body weight reduction technologies from other technologies, as we assume these will not be utilized under weight class regulation. Next, we evaluate the cost-effectiveness of possible combinations of technologies. We then estimate cost functions based on the most cost-effective combinations.

3.5.1 Survey of fuel efficiency technologies

Some surveys in the literature have considered improvement rates and the costs of fuel efficiency technologies. We utilize the data in ARB (Air Resource Boards, 2004) and Smokers *et al.* (2006), where the assumed vehicle conditions are comparable with the Japanese situation.

The data in ARB correspond to small- and ordinary-sized vehicles in Japan. We applied the data for small-size vehicles to mini-sized vehicles as well. The data in Smokers *et al.* correspond to mini-, small-, and ordinary-sized vehicles, though the base vehicles on which these data are based have manual transmissions, which are rarely used for private passenger cars in Japan. Therefore, we cannot directly use the transmission data in this work. We do utilize the weight reduction and hybrid technologies data from this work. Furthermore, while the cost of automated manual transmission is assumed to be zero in the ARB work, we assume that this requires dual-clutch technologies, whose cost data we derive from Owen and Gordon (2002).

Table 3.2 shows a compilation of data on fuel consumption reduction rates and retail prices in US dollars by vehicle size. In this table, the technologies are categorized into engine, transmission, accessory and body technologies. Here, the reduction rate is based on L/km fuel consumption, and the costs are assumed to be the retail prices at the mass production stage. The shaded boxes refer to weight reduction technologies.

Table 3.2 Fuel consumption reduction rates and technology costs

		Fuel consumption reduction rate			Retail price equivalent ($)		
Category	**Technology**	**mini**	**small**	**ordinary**	**mini**	**small**	**ordinary**
Engine	VVT	0.03	0.03	0.04	98	98	196
	DVVL	0.04	0.04	0.04	203	203	357
	CVVL	0.05	0.05	0.06	280	280	581
	Camless	0.11	0.11	0.11	676	676	764
	DAC	0.00	0.00	0.06	0	0	183
	GDI-S	0.00	0.00	0.01	189	189	259
	GDI-L	0.06	0.06	0.09	728	728	959
	gHCCI	0.04	0.04	0.06	560	560	840
	Turbo-M	0.09	0.10	0.10	293	391	489
	Turbo-S	0.12	0.12	0.12	509	587	665
Transmission	5AT	0.02	0.02	0.01	140	140	140
	6AT	0.03	0.03	0.03	70	70	105
	AMT	0.08	0.08	0.07	183	313	574
	CVT	0.04	0.04	0.03	210	210	245
Accessory	EPS	0.01	0.01	0.01	20	20	39
	ALT	0.01	0.01	0.01	56	56	56
	42V+IS	0.07	0.07	0.04	609	609	609
	42V+MA	0.10	0.10	0.06	902	902	902
	mod-HEV	0.18	0.18	0.18	1,565	2,087	2,609
	adv-HEV	0.36	0.36	0.36	3,652	4,565	5,478
Body	mild-WR	0.01	0.01	0.01	29	37	44
	med-WR	0.02	0.02	0.02	74	117	150
	str-WR	0.06	0.06	0.05	277	383	545

Notes:
VVT: Variable Valve Timing, **DVVL**: Discrete Variable Valve Lift, **CVVL**: Continuous VVL
Camless: Electromagnetic Camless Valve Actuation, **DAC**: Cylinder Deactivation
GDI-S: Gasoline Direct Injection–Stoichiometric, **GDI-L**: GDI-Lean-Burn Stratified
gHCCI: Gasoline Homogeneous Compression Ignition
Turbo-M: Medium downsizing with turbocharging, **Turbo-S**: Strong downsizing with turbocharging
AMT: Automated Manual Transmission, **CVT**: Continuously Variable Transmission
EPS: Electric Power Steering, **ALT**: Improved Alternator
42V+IS: 42-Volt 10 kW ISG (Start Stop), **42V+MA**: 42-Volt 10 kW ISG (Motor Assist)
mod-HEV: Moderate Hybrid-Electric Vehicle, **adv-HEV**: Advanced HEV
mild-WR: Mild Weight Reduction, **med-WR**: Medium WR, **str-WR**: Strong WR

Source: Compiled by authors based on Air Resource Boards (2004) and Smokers *et al.* (2006).

3.5.2 Estimates of cost functions

The transmission and body technologies in Table 3.2 cannot be combined with other technologies in their respective categories. On the other hand, some technologies in the engine and accessory categories can be used together. We examined technologically possible combinations for each category; in the category of engine technologies, there are 62 and 22 possible combinations with and without weight reduction technologies, respectively. The category of accessory technologies does not include weight reduction technologies, so there are 14 combinations of technologies in this category. All entries in the body technology category are weight reduction technologies.

Based on this set-up, there are in total 17,360 and 1,540 combinations of technologies in all categories with and without weight reduction technologies, respectively. The number of possible combinations is quite different between these two cases, which implies that weight class regulations cut possible technology options and may increase the cost of achieving the target. The fuel consumption reduction rate and the change in price are estimated based on the following equations:

$$\alpha_K = 1 - \prod_{k \in K} (1 - \alpha_k), \tag{3.5}$$

$$c_K = \sum_{k \in K} c_k \tag{3.6}$$

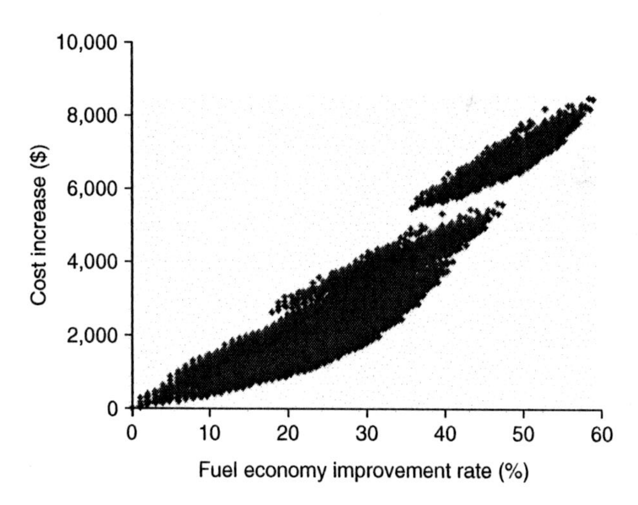

Figure 3.3 Price increase versus fuel consumption reduction rate (with weight reduction technologies, ordinary-sized car)

Here α_k is the fuel consumption reduction rate for technology k in the combination K and c_k is its Retail Price Equivalents (RPE). Figure 3.3 shows a plot of RPE versus reduction rate for all combinations of technologies for ordinary-sized cars with weight reduction technologies. In this figure, the lower right points have higher reduction rates and lower prices; that is, they are cost-effective. The envelope curve of cost-effective combinations seems to have an inflexion point at around a fuel consumption reduction rate of 30 per cent, and the gradient of the price increases thereafter.

To estimate cost functions, we segregate the combination data into 1 per cent segments. We keep data for the lowest 30 per cent within each segment, and estimate regression curves based on the data retained. These curves are then fitted to fourth-order polynomials. The fit curves

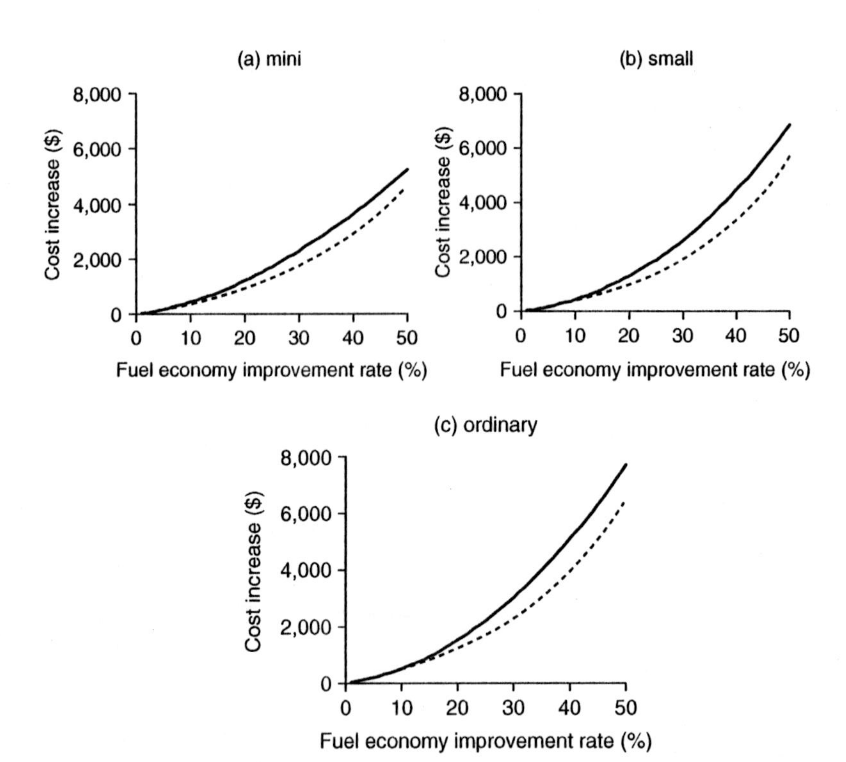

Figure 3.4 Estimated cost functions

Notes: Solid line: with weight reduction technologies; dotted line: without weight reduction technologies.

Table 3.3 Estimated cost function

	Without weight reduction			With weight reduction		
	mini	small	ordinary	mini	small	ordinary
a1	2,691	2,361	2,785	2,808	3,433	4,512
a2	17,023	20,675	24,173	8,265	6,302	6,796
a4	5,804	8,482	4,324	19,362	38,895	41,095

Note: Cost function: $C=a1\times x+a2\times x^2+a4\times x^4$ (x: fuel economy improvement rate).

are shown in Figure 3.4, and the fitting parameters are shown in Table 3.3.

These figures show that weight reduction technologies do not affect the cost at reduction rates below 15 per cent, but that weight reduction technologies do push down cost increases for reduction rates above 15 per cent. As far as vehicle size is concerned, heavier vehicles require higher costs to achieve improvements in fuel economy. These patterns are consistent with our experience and with the data shown in Table 3.2.

3.6 Impact of regulations on vehicle price

In this section, we analyze the impact of two regulatory schemes – the weight class approach and the CAFE approach – in the case of the Japanese market, using the cost functions estimated above. This analysis requires, as a baseline, an estimate of the current rate of penetration of fuel efficiency technologies. We must also make assumptions regarding fluctuations in sales distribution by weight class in order to estimate the impact of the CAFE regulation.

In what follows, we first estimate the rate of penetration of fuel efficiency technologies, and their contribution to improvements in fuel economy, for the specific case of the Japanese market in 2004. We next analyze changes in the market share commanded by three vehicle weight classes, based on vehicle registration statistics. Then, based on this information, we estimate changes in the vehicle price necessary to achieve the Japanese 2015 standard, for the cases of both CAFE and weight class regulations. Finally, we discuss the sensitivity of vehicle price to standard levels under the two regulatory approaches.

3.6.1 Estimates of the rate of penetration of fuel efficiency technologies

In order to estimate the increase in vehicle price caused by the introduction of fuel efficiency technologies, we must first understand the current status of fuel economy improvements due to technologies already installed. Based on data from the Japanese Ministry of Land, Infrastructure, Transport and Tourism (JMLIT), we estimate the rate of penetration in 2004 of nine technologies: Variable Valve Timing (VVT), Cylinder Deactivation (DAC), Gasoline Direct Injection–Stoichiometric (GDI-S), GDI-Lean-Burn Stratified (GDI-L), Continuously Variable Transmission (CVT), Idling stop (IS), Hybrid-Electric Vehicle (HEV), and Turbocharging. The results are shown in Table 3.4.

Variable valve timing penetrated into all classes, and particularly into the ordinary-sized class: over 90 per cent of ordinary-sized vehicles had this technology installed. While GDI-S was installed in 15 per cent of ordinary-class vehicles, penetration rates for this technology are very low in the other classes. For the small-sized class, the penetration rate of CVT is more than 30 per cent, although the penetration rate for this technology is rather low for the ordinary-sized class due to the

Table 3.4 Penetration rates for fuel efficiency technologies in 2004

	VVT	DAC	GDI-S	GDI-L	CVT	IS	HEV	Turbo
mini	50%	0%	1%	0%	4%	0%	0%	18%
small	87%	0%	5%	2%	32%	2%	2%	2%
ordinary	91%	2%	15%	3%	7%	1%	1%	0%

Table 3.5 Average fuel efficiency improvement due to technologies in 2004

	Fuel consumption in 2004 (L/100km)		
	Average	**Baseline**	**Improvement rate**
mini	5.33	5.52	3.4%
small	6.19	6.54	5.3%
ordinary	9.64	10.20	5.6%
Average	6.91	7.28	5.0%

Baseline: disregarding the effect of fuel efficiency technologies.

durability problem. Turbocharger penetration is high in the mini-sized class. Penetration rates for other technologies are quite low.

Based on these penetration rates, and using Eq. (3.5), fuel economy improvement rates for 2004 are estimated as shown in Table 3.5. On average, these technologies have improved fuel economy about 5 per cent over the baseline vehicle. In the real market, various other technologies, such as friction reduction, are utilized, and the actual improvement over 1995 is much higher. We assume these other technologies are installed on all vehicles without price increase, and, in estimates of future improvement, we take the improvement rates in Table 3.5 as initial values, and their costs are already reflected in the vehicle prices.

Applying the cost functions derived in Section 3.5, we estimate that fuel efficiency technologies have increased vehicle prices by $100 for mini-class vehicles, $200 for small-class vehicles, and $250 for ordinary-class vehicles.

3.6.2 Fluctuations in sales distribution by weight class

To assess the impact of CAFE regulations, we need to make assumptions regarding fluctuations in sales distribution by weight class. The regulations will remain in the future, but the sales distribution will shift due to market factors, such as consumer preferences, which are not directly controlled by auto-makers. Thus we assume that the distribution will change stochastically, and that auto-makers will improve the

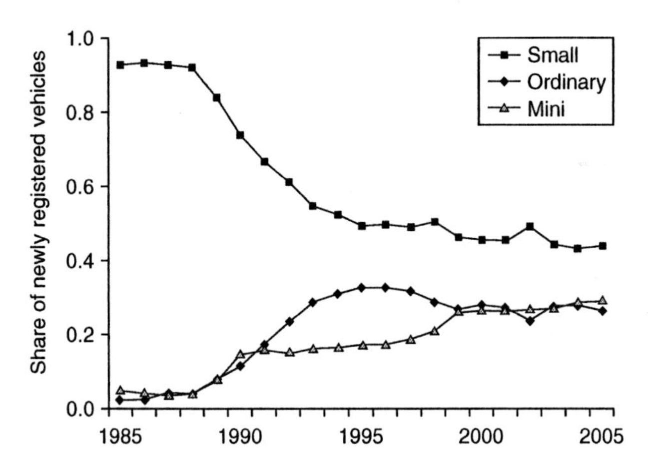

Figure 3.5 Trend in sales distribution by weight class
Source: Japan Automobile Dealers Association (2010).

Table 3.6 Estimated parameters of probability density functions

	mini	small	ordinary
Mean (share in 2004)	0.29	0.43	0.28
Standard deviation (1985–2005)	0.08	0.19	0.11

fuel efficiency of their vehicles to control for the risk of non-compliance with regulations with a certain probability.

Figure 3.5 shows the trend in the sales distribution by weight class since 1985. In the early 1990s, the distribution shifted significantly, reflecting revisions to the automobile tax system at the end of the 1980s. If the share of mini-class cars decreased and that of ordinary-class cars increased, the risk of non-compliance with the CAFE standards would increase. On the other hand, if the share of mini-class cars increased and that of ordinary-class cars decreased, the risk of non-compliance would decrease. If we expect the standard not to be loosened, but rather to be tightened, during the next 20 years, then auto-makers may introduce technologies based on a consideration of shifts in the sales distribution over the past 20 years. Possible changes here include not only a recovery of market share for small-class vehicles, but also an increase in share for the ordinary class and a decrease in share for the mini class.

The market share for each class is assumed to be a normally distributed random variable, with mean value set equal to the actual market share value in 2004 and standard deviation set equal to the variance in the actual market share data from 1985 to 2005. The parameters for each weight class are shown in Table 3.6.

However, since the sum of the shares must be unity, the probability density function in Eq. (3.4) must be formulated as a joint probability density with a total constraint. Specifically, using the symbol **RC** to denote the set of market share distribution vectors that satisfy the two constraints that every share is greater than zero and the sum of all shares is 1, the joint probability is given as follows:

$$p(\mathbf{r}) = \prod_g p_g\left(r_g\right) \Big/ \int_{\mathrm{RC}} \prod_g p_g\left(r_g\right) d\mathbf{r} \tag{3.7}$$

where p_g is the probability density for class g to have market share r_g.

3.6.3 Impact of regulations on vehicle price

In this section, we estimate prices for vehicles designed to meet Japanese standard levels in 2015, both for weight class regulations and for CAFE

Table 3.7 Fuel economy improvement rates necessary to achieve the 2015 target

	Fuel consumption in 2004 (L/100km)		2015 standard (L/100km)	Required improvement rate	
	Baseline	**Average**		**vs. baseline**	**vs. 2004 average**
mini	5.52	5.33	4.33	22%	19%
small	6.54	6.19	4.96	24%	20%
ordinary	10.20	9.64	7.31	28%	24%
Average	7.28	6.91	5.44	25%	21%

regulations. The standard for the CAFE approach is defined by the harmonic average of the 2015 standards, assuming the distribution of sales by weight class to be fixed at the actual distribution in 2004.

We first aggregate the 2015 standards into three classes – mini, small, and ordinary– based on the report by METI (Ministry of Economy, Trade and Industry) in Japan, and estimate the required improvement rate from the baseline fuel economy data listed in Table 3.5. The required improvement rates are shown in Table 3.7. In this table, the column of 2015 standards shows the targets by weight class, and the column marked 'Average' is the target for the CAFE.

The required improvement rates by weight class indicate that heavier classes need larger improvements. On average, the improvement rate is 25 per cent over the baseline vehicle and 21 per cent over the 2004 average vehicle. For the case of weight class regulations, we may use the required improvement rate and the estimated cost function to calculate immediately an estimated increase in vehicle price: $1,250 for mini-class vehicles, $1,630 for small-class vehicles, and $2,530 for ordinary-class vehicles. On average, this is an increase of $1,770 over the 2004 price.

For the case of CAFE regulations, the required improvement rate by weight class is obtained by solving Eq. (3.4). If auto-makers are assumed to set the probability of compliance with the standard at 95 per cent, then the vehicle price increase is estimated to be $1,130 for mini-class vehicles, $1,860 for small-class vehicles, $2,490 for ordinary-class vehicles, and $1,830 on average. As shown in Figure 3.4, weight reduction technologies are available under the CAFE approach, resulting in lower price increases as compared with the weight class regulatory approach. However, auto-makers must be more aggressive in introducing fuel efficiency technologies to control risks due to shifts in the market share distribution. As a result, price increases in the CAFE approach are higher, on average, than those in the weight class regulation approach.

Meanwhile, if auto-makers assumed a compliance probability of just 90 per cent, the average vehicle price increase would be $1,690, which is lower than that in the weight class regulation approach. This is due to the alleviation of the constraint condition in Eq. (3.4), which implies that the awareness of auto-makers of the risks of non-compliance affects the rate of technology penetration as well as the vehicle price. Figure 3.6 shows the sensitivity of the average vehicle price to the assumed probability of compliance, indicating that vehicle price under the CAFE regulations can be either higher or lower than that under weight class regulations, depending on the probability of compliance assumed by auto-makers.

The probability of compliance is believed to be set based on damages suffered by auto-makers upon failure to achieve compliance with the standard. If consumers have considerable interest in energy-saving technologies, then non-compliance may result in negative impressions attached to vehicles made by the auto-makers concerned. In this case, the probability of compliance chosen by auto-makers will be high enough to avoid such negative consequences. On the other hand, if consumers have less interest in energy-saving technologies, then

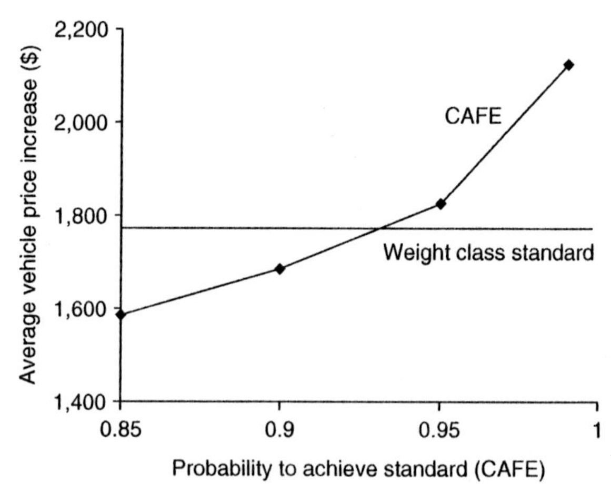

Figure 3.6 Vehicle price increase versus assumed probability of non-compliance

Table 3.8 Fuel economy improvement rates and vehicle price increases for CAFE and weight class regulations

		Weight class standards	CAFE (achievement probability)			
			(99%)	(95%)	(90%)	(85%)
Rate of improvement over 2004 average	mini	19%	25%	22%	20%	21%
	small	20%	30%	27%	26%	23%
	ordinary	24%	31%	29%	29%	28%
	average	21%	29%	27%	26%	25%
Vehicle price increase ($)	mini	1,250	1,450	1,130	1,040	1,100
	small	1,630	2,220	1,860	1,670	1,440
	ordinary	2,530	2,670	2,490	2,380	2,310
	average	1,770	2,130	1,830	1,690	1,590

auto-makers will choose a lower target probability. Today, the majority of consumers are expected to pay significant attention to fuel economy, in view of the present global situation of high oil prices and concerns over global warming. This causes auto-makers to set higher targets for probability of compliance. The existence of auto-makers whose average fuel economy far exceeds the CAFE standard, as shown in Figure 3.2, can be taken as evidence for this view.

Table 3.8 summarizes the results of the above analysis and shows rates of fuel economy improvement as well as vehicle price increases, as compared with 2004 averages under both weight class and CAFE approaches. Both approaches have the same standard levels on average, but the CAFE induces higher fuel economy improvement than the weight class approach in all cases. This reflects the risk-averting behavior of auto-makers considering changes in the sales distribution by vehicle class. On the other hand, the increase in vehicle price needed to attain a certain level of fuel economy improvement is lower in the CAFE, reflecting the availability of weight reduction technologies.

3.7 Sensitivity of vehicle prices to standard levels

As discussed above, at the standard levels set by Japanese regulations for 2015, vehicle price increases under the CAFE approach can be either higher or lower than those under the weight class approach, depending on the compliance probability chosen by the auto-maker. In this section, we analyze the impact of standard levels on vehicle prices and the differences among the regulatory approaches.

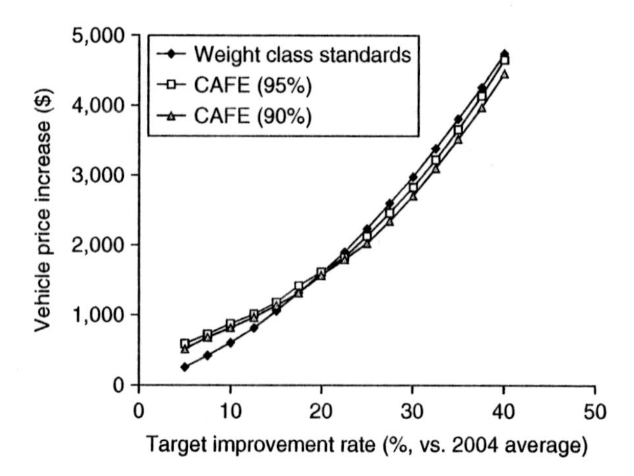

Figure 3.7 Vehicle price versus improvement rate

Figure 3.7 shows the change in vehicle price versus target improvement rate relative to the 2004 average over the range of 5 per cent to 40 per cent for both approaches. For the CAFE approach we plot two lines, for compliance probabilities of 90 and 95 per cent. For the weight class regulation approach, the price plotted is an average price derived from the weighted average by sales share.

This figure demonstrates that vehicle prices under weight class regulation are lower than those under the CAFE when the standard levels are low, but that the CAFE gives lower prices when standard levels are high. Although the availability of weight reduction technologies does not affect vehicle prices at low improvement levels, as shown in Figure 3.5, the CAFE requires additional improvements beyond the standard target to avoid the risk of non-compliance. On the other hand, the availability of weight reduction technologies affects the vehicle price at higher improvement levels, which would make vehicle prices in the CAFE approach lower than those in the weight class regulation approach.

As seen in Table 3.7, the Japanese standard for 2015 amounts to a 21 per cent improvement over the 2004 average fuel economy. The figure shows that the vehicle price is almost the same under both approaches – around a 20 per cent improvement. From this result, we can conclude that weight class regulations are more effective than the CAFE in Japan, and will be at least until the 2015 standard levels are set. However, if more stringent standards are required after 2015, the analysis suggests that the CAFE should also be examined as a potentially effective regulatory approach.

3.8 Benefits of automotive fuel economy regulations for consumers

With the advent of fuel-efficient technology options such as gas–electric hybrid vehicles, one of the most crucial concerns for manufacturers will be consumer acceptance of increased vehicle prices due to the efforts of manufacturers to meet fuel economy regulations. In the European countries and the US, the economic and social impact of fuel economy targets is generally assessed during the rule-making process. The Transport Research Board and the National Research Council (2002) have issued a comprehensive study of the impact of the Corporate Average Fuel Economy (CAFE) standards, which resulted in consumers realizing net savings from improved fuel efficiency. The National Center for Statistics and Analysis (2005) has also provided a detailed study of the impact of CAFE on light trucks. Several supporting studies regarding the development of strategies to achieve the 120g CO_2/km target for passenger cars have been conducted in Europe; these studies also include the socio-economic impact of fuel-efficient technologies (Bates *et al.*, 2000; Brink *et al.*, 2005). Studies in Japan, on the other hand, have largely been limited to technological and mass-production feasibility, with discussion of quantitative economic impact kept to a minimum due to a lack of reliable information on costs and benefits.

The purpose of this section is to evaluate the impact of the 2015 Japanese fuel economy standard on net consumer savings. This section is organized as follows. First, we formulate the problem of setting fuel economy targets in such a way as to maximize net consumer savings under weight class regulations. Next, we apply the cost curves obtained in previous sections to estimate net consumer savings realized by fuel economy improvements, which in turn allows us to derive the optimal target levels that maximize consumer savings. We also analyze the effect of each IW class on fuel economy. We then compare our estimated optimal targets with the actual 2015 standards in order to assess the fundamental economic soundness of the regulations. Finally, we consider the potential impact of some factors neglected by our quantitative analysis, and we discuss some practical issues as well as some points requiring further study.

3.8.1 Assessing the impact of regulations on fuel economy

We assume the following conditions in this analysis: a) the portion of total sales occupied by each separate IW class is fixed at its actual observed value for 2004; b) all technologies and features other than fuel economy are constant; c) the entire cost increase caused by fuel economy improvements is applied to the vehicle price; and d) there is no difference in the increase of production costs among manufacturers. These assumptions are too restrictive to allow us to analyze the effects of the regulations on market demand or on individual manufacturers; however, these assumptions simplify and clarify the analysis needed to estimate net consumer savings.

As discussed in Greene and Hopson (2003), the net consumer savings resulting from fuel economy improvements are defined as the difference between the net fuel cost savings over the vehicle's lifetime and the retail price increase of the vehicle. The net consumer cost savings can thus be expressed as follows:

$$\Pi(\boldsymbol{\alpha}^{15}) = L\, p_f\, \Gamma\, \eta \sum_g \{s_g\, e_g^{00}\, \alpha_g^{15}\} - (1+\tau)\sum_g s_g\, (f_g(\alpha_g^{15})). \quad (3.8)$$

where Π denotes net consumer savings, $\boldsymbol{\alpha}^{15}$ ($=\{\alpha_g{}^{15} \mid g \in G\}$) is a vector whose components are the rates of fuel consumption reduction (as compared to baseline fuel consumption) for each IW class (the index g runs over all IW classes), L is the annual travel distance per vehicle, p_f is the price of fuel, Γ is a constant reflecting the fuel saving discount during the vehicle's lifetime, η is the ratio of real fuel consumption to test fuel consumption, s_g is the market share fraction for IW class g, $e_g{}^{00}$ is the baseline (for year 2000) fuel consumption in l/km, η is the tax on the vehicle purchase, and f_g is the retail price increase due to the fuel consumption improvement over the baseline obtained in Section 3.7. Γ can be derived from the annual discount rate and the vehicle lifetime according to the relation:

$$\Gamma = \frac{(1+r)^{T+1} - 1}{r \cdot (1+r)^T}, \quad (3.9)$$

where r is the annual discount rate and T is the vehicle lifetime.

The improvement rates that maximize the average net consumer savings are derived from the following formula:

$$\max_{\boldsymbol{\alpha}^{15}} \Pi(\alpha^{15}),$$

$$\text{s.t. } \bar{\alpha}^{15} = (\Sigma s_g \cdot e_g \cdot \alpha_g^{15}) / \Sigma s_g \cdot e_g,$$

$$\alpha_g^{15} \geq \alpha_g^{04} \text{ for } \forall g. \quad (3.10)$$

Here $\bar{\alpha}^{15}$ is the average improvement target and α_g^{04} is the improvement in 2004 over the baseline shown in the fourth column of Table 3.5.

We assume the following values for the variable parameters in these equations: the annual travelΓ length l = 9,800 km; the fuel price p_f = \$1.64/l; the fraction of real fuel consumption η= 1.4; the vehicle purchase tax τ= 10 per cent; the discount rate r = 7 per cent; and the vehicle lifetime T = 10 years. Values for the other variables, such as the market share fraction for each IW class and the baseline fuel consumption, are listed in Section 3.7. We refer to the estimates derived under these conditions as the 'base scenario' in the remainder of this section.

The solution of the above equation yields net consumer savings for a given average improvement target. We considered 69 values of the average improvement target, ranging from 6 per cent to 40 per cent in 0.5 per cent steps. The minimum target value was derived based on the actual improvement rate in 2004, as shown in Table 3.5, and the second constraint in the formula.

To account for uncertainties in estimated costs, we also estimated scenarios in which vehicle price changes deviated by ±15 per cent from the original curve. With the change in price denoted by μ, the retail price increase function is:

$$f_g = \mu \cdot f_g^{EC} \text{ for } \forall g. \tag{3.11}$$

where f_g^{EC} is the original price function.

3.8.2 Benefits to consumers

Figure 3.8 shows estimated net consumer savings (over baseline) versus fuel consumption reduction rate, both for the base-cost scenario and for the ±15 per cent vehicle-price scenarios. As shown in Table 3.5, the fuel consumption rate had already decreased by approximately 5 per cent in 2004. This resulted in net consumer savings of approximately \$310 in the base-cost scenario. We estimate that the maximum net savings in this scenario occur at a fuel consumption reduction rate of 17 per cent. Figure 3.8 also shows that, when technology prices are higher, both net consumer savings and optimal reduction rates will be lower. For the ±15 per cent vehicle-price scenarios, the optimal reduction rate lies within the range of 14 to 22 per cent; the optimal reduction rate in average fuel consumption from year 2004 was in the 10 per cent to 17 per cent range.

Net consumer savings also depend on fuel price. Figure 3.9 shows scenarios in which the price of fuel deviates ±15 per cent from the base scenario. Higher fuel prices, which increase the value of fuel savings,

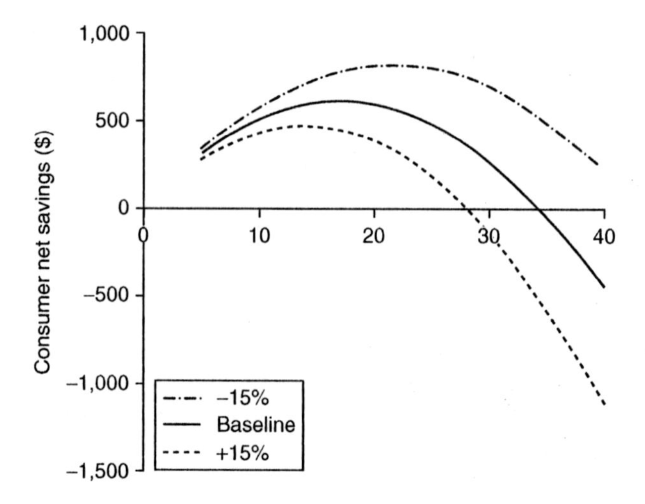

Figure 3.8 Net consumer savings versus fuel consumption reduction rate and the effect of vehicle price

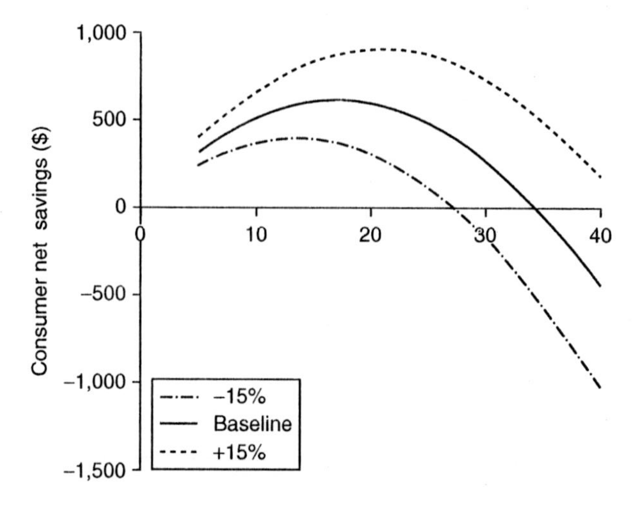

Figure 3.9 Net consumer savings versus fuel consumption reduction rate and the effect of gasoline price

result in higher net consumer savings and a higher optimal value of the fuel consumption reduction rate. A comparison of vehicle prices reveals that net savings decline sharply after the reduction rate exceeds the optimal point.

The absolute value of the elasticity in the maximum net savings is 2.3 with respect to fuel price and 1.9 with respect to vehicle price, while the elasticities in the optimal fuel consumption reduction rate are almost identical with respect to fuel and vehicle price – 0.14 and 0.15, respectively. This result implies that changes in the price of fuel affect net consumer savings much more than do changes in vehicle price, whereas changes in the two prices have nearly the same impact on the optimal reduction rate. Recently, a shift in demand toward fuel-efficient vehicles has been observed as a result of surging oil prices. The above result implies that technology cost reductions may have a similar impact on the optimal reduction rate.

Figure 3.10 shows, for each IW class, 1) the optimal target (as determined above), 2) the actual targets for the years 2010 and 2015, 3) the actual average fuel consumption for 2004, and 4) the actual fuel consumption of the top-runner vehicle in 2004. The 'top-runner' vehicle is defined to be the most fuel-efficient of all vehicles to have sold 1,000 units or more, excluding manual transmission and hybrid vehicles. The figure illustrates that the actual average fuel consumption levels for 2004 were almost on a par with the 2010 targets. For weight classes lighter than IW2000, the 2004 top-runners may potentially be so fuel-efficient that their fuel consumption reduction rates even exceed the optimal levels estimated above, but the fuel economy levels of the top-

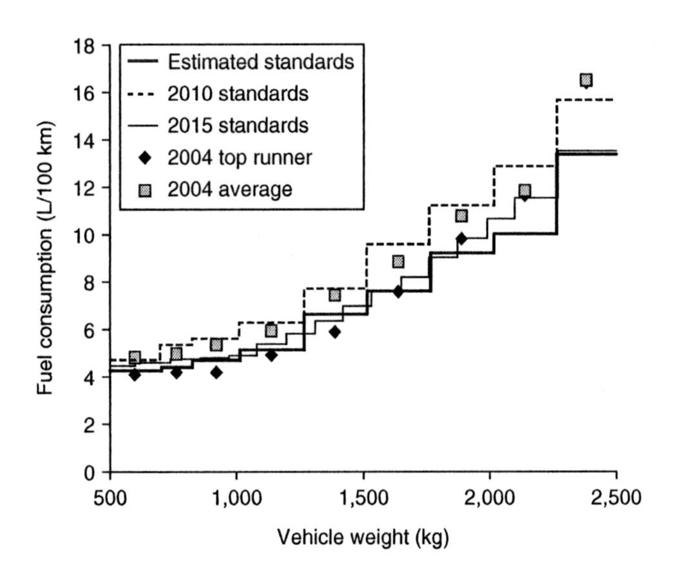

Figure 3.10 Fuel consumption targets by IW class

runners in the other IW classes are likely to fall below the optimal targets. For the heaviest classes, and particularly for the classes IW2000 to IW2500, levels of fuel economy much higher than those of the top-runners would be necessary to maximize net consumer savings.

It is clear that the actual 2015 target levels are in fact quite similar to the estimated optimal targets, albeit with different segmentations by vehicle weight class. However, there are still certain other policy issues left to consider, which we will discuss in Section 3.9.

3.9 Conclusions

In this chapter we compared the impact of two regulatory schemes, the CAFE and weight class standards, on vehicle prices. We next analyzed the consumer benefit under the weight class standards based on the cost analysis for the Japanese standard in 2015. Our main findings can be summarized as follows.

In our analysis of the costs of fuel efficiency technologies, we found that marginal vehicle price increases versus improvement rate, and the availability of weight reduction technologies, do not affect prices at lower improvement levels, but do make a difference at higher levels.

We found that possible shifts in sales distribution by weight class would pose a risk for non-compliance with the CAFE regulations, but not for the weight class regulations. In the CAFE approach, the automakers' attempts to control these risks would promote aggressive technology installation and would increase vehicle prices as compared with the weight class regulation approach. This behavior was assumed to depend not only on the penalties for non-compliance with the standard, but also on consumer preferences for fuel-efficient vehicles.

Based on these findings, we compared the impacts of both approaches on vehicle price. We demonstrated that weight class regulations give lower vehicle prices when standard levels are low, but that the CAFE approach may result in lower vehicle prices when standard levels grow high enough. For the Japanese standard in 2015, the two regulatory approaches were estimated to have nearly the same impact on vehicle price. If standards are to be strengthened in Japan after 2015, the CAFE approach could be considered as an alternative approach.

As to consumers' benefit analysis, consumer benefit, as a function of fuel economy improvement rate, was found to have optimal points; in other words, certain levels of fuel economy regulation maximize net consumer benefit, whereas standards set too high have negative effects. We estimated that the standard levels set by the 2015 Japanese regulations will have a positive impact on consumer benefit. Based on the sensitivity analysis, we found that the cost of automotive technology, and the cost of fuel, had considerable effect on consumer benefit. We estimated that the impacts of these two factors on consumer acceptance of fuel-efficient technologies are almost identical, at least from an economic viewpoint.

It should be noted that the analysis in this chapter depends on various assumptions and limited information. For the estimated cost functions, the technology costs obtained from the literature in the US and Europe might not fully reflect the Japanese situation. For the analysis of risks under the CAFE, the target probability of compliance and the probability of fluctuations in market share were estimated as national averages, but in practice would differ among auto-makers. Regarding consumer benefit, our analysis has been limited to matters lying within the narrow scope of economic benefits. More general assessments will require an analysis of social benefits. This would include studies of the value of reduced greenhouse gas emissions and of improved national energy security, although methods for quantifying such matters are not yet established.

In addition to these analytical limitations, to discuss the feasibility of the CAFE approach as a real policy option, we would need to elaborate on differences among auto-makers. As the original CAFE standard was specified in a single, uniform way, differences in sales distributions by vehicle weight among auto-makers affected the equity of competitive conditions in the market. This observation underlies the recent modification of CAFE to reflect vehicle footprint; we note as well that proposed regulations in the EU also include standards adjusted by vehicle weight. A study of new approaches that takes into account the merits of both schemes – namely, the equity of competitive conditions and the unlimited availability of weight reduction technologies – is needed. Finally, analytical comparisons with other regulatory approaches – such as approaches that combine regulations with tax and/or subsidy structures – are a necessary input for a more wide-ranging discussion of future enhancements to regulatory structures. We will discuss this issue in detail in Chapter 4.

References

Air Resources Board (2004) *Staff Proposal Regarding the Maximum Feasible and Cost-Effective Reduction of Greenhouse Gas Emissions from Motor Vehicles*, Sacramento: California Environmental Protection Agency.

An, F. and Sauer, A. (2004) *Comparison of Passenger Vehicle Fuel Economy and Greenhouse Gas Emission Standards around the World*, Arlington, TX: Pew Center on Global Climate Change.

An, F., Gordon, D., He, H., Kodjak, D., and Rutherford, D. (2007) *Passenger Vehicle Greenhouse Gas and Fuel Economy Standards: A Global Update*, Washington, DC: International Council on Clean Transportation.

Bates, J., Brand, C., Davison, P., and Hill, N. (2000) *Economic Evaluation of Emissions Reductions in the Transport Sector of the EU, AEA Technology Environment for DG Environment*, Brussels: European Commission.

Brink, P., Skinner, I., Fergusson, M., Haines, D., Smokers, R., Burgwal, E., Gense, R., Wells, P., and Nieuwenhuis, P. (2005) *Service Contract to Carry out Economic Analysis and Business Impact Assessment of CO2 Emissions Reduction Measures in the Automotive Sector: Final Report*, Publication B4-3040/2003/366487/MAR/C2, London: UK Institute for European Environmental Policy.

European Commission (2010) *Reducing CO2 Emissions from Light-Duty Vehicles*, available at http://ec.europa.eu/environment/air/transport/co2/co2_home.htm (accessed November 11, 2010).

Greene, D. L. and Hopson, J. L. (2003) 'An Analysis of Alternative Forms of Automotive Fuel Economy Standards for the United States', *Transportation Research Record: Journal of the Transportation Research Board*, 1842: 20–8, Washington, DC: National Research Council.

International Council on Clean Transportation (2010) 'U.S. Light-Duty Vehicle GHG and CAFE Standards: Final Rule Summary', *Policy Update*, 7, available at http://www.theicct.org/documents/0000/1438/ICCTpolicyupdate7.pdf (accessed July 27, 2010).

National Center for Statistics and Analysis (2005) *Corporate Average Fuel Economy and CAFE Reform for MY 2008–2011 Light Trucks*. Washington, DC: US Department of Transportation.

National Highway Traffic Safety Administration (2006) *Corporate Average Fuel Economy and CAFE Reform for MY 2008–2011 Light Trucks*. Washington, DC: US Department of Transportation.

Owen, N. and Gordon, R. (2002) *'Carbon to Hydrogen' Roadmaps for Passenger Cars: A Study for the Department for Transport and the Department of Trade and Industry*, Publication RD. 02/3280, London: UK Department for Transport.

Plotkin, S., Greene, D. L., and Duleep, K. G. (2002) *Examining the Potential for Voluntary Fuel Economy Standards in the United States and Canada*. Publication ANL/ESD/02-5. Argonne, IL: Center for Transportation Research, Argonne National Laboratory, US Department of Energy.

Smokers, R., Vermeulen, R., Mieghem, R., Gense, R., Skinner, I., Fergusson, M., MacKay, E., Brink, P., Fontaras, G., and Samaras, Z. (2006) *Review and Analysis of the Reduction Potential and Costs of Technological and other Measures to Reduce CO2-Emissions from Passenger Cars*, TNO Report 06.OR.PT.040.2/RSM, Report to the European Commission.

Transportation Research Board and the National Research Council (2002) *Effectiveness and Impact of Corporate Average Fuel Economy (CAFE) Standards*, Washington, DC: National Academy Press.

4

Future Directions for Fuel Efficiency Policy: Evolving from Fuel Efficiency Standards toward Indirect Regulations

Hiroaki Miyoshi, Masayuki Sano, Masanobu Kii, and Yuko Akune

4.1 Introduction

In Chapter 2, we reviewed the trajectory of automotive technology advances – and the evolution of government technology policies – in post-war Japan, and we anticipated future directions for automotive technology innovation. Our conclusion was that, although we should consider expanding the use of economic measures, nonetheless the best way to guarantee the safety and reliability of future automobiles is to *preserve* and continue the existing regulatory structure in essentially its present form. In this chapter, we consider responses to the problem of global warming – a phenomenon which poses an urgent threat to the continued existence of human civilization – and reach a dramatically different conclusion, namely, that direct government regulation of fuel efficiency, the primary policy mechanism that has been used to address global warming in the past, is a fundamentally *inappropriate* response to the critical problem of climate change. We begin in Section 4.2 with an introduction to the process by which automotive regulatory policy has traditionally been crafted in Japan; we note that the content of regulations, and the standard levels they mandate, have traditionally been fixed through tight-knit collaboration between industry, academia, and government, and have been designed to minimize the likelihood of any car-maker failing to meet regulatory standards or being forced to withdraw from particular markets. Next, in Section 4.3, we survey Japan's hierarchy of automobile taxes, which for many years has functioned as a dedicated funding source for the construction and maintenance of roads; we examine the appropriateness of the fee structure of Japan's automobile tax system when thought of as a type of public utility charge, and we point out certain aspects of Japan's automobile tax system that are problematic from the perspective of environmental friendliness. Finally, in Section 4.4, after a brief summary of the current state of the global warming phenomenon, we emphasize the importance of two critical directions for future government policy: (1) a shift in policy away from *direct* regulation of fuel efficiency and toward *indirect* regulation (economic regulations), and (2) a reform of the automobile tax system.

4.2 The structure of Japan's automotive technology policies

In this section, we discuss the process by which automobile-related regulatory policy has traditionally been crafted in Japan.

4.2.1 Automotive technology policies focusing on regulations

As discussed in Chapter 2, automotive technology policies in post-war Japan have focused on regulations, with the administration of policies managed by four authorities: the Ministry of Land, Infrastructure, Transport and Tourism (which administers the Road Transport Vehicle Act, the legislation creating the Inspection System that forms the basis for the implementation of regulatory systems), the Ministry of Economy, Trade and Industry, the Ministry of the Environment, and the National Police Agency. The planning, analysis, and decision-making required to draft the specific content of policies and regulations are carried out by deliberative bodies, including councils, inquiry panels and committees, with the ministry responsible for administering each law serving as the secretariat for deliberations over that law. Tables 4.1 and 4.2 list the main pieces of legislation addressing automotive technological policy in Japan, together with the authorities in charge of overseeing each component.

A consequence of this process for constructing regulation-based automotive technology policies is that decisions are made in close collaboration among government, academia, and industry. The Japan Automobile Manufacturers' Association (JAMA), an industry organization of 14 Japanese car-makers,[1] plays an important role in this collaboration.

Several committees (*iinkai*) have been established within JAMA to address various challenges facing the automobile industry. Within these committees, subcommittees (*bukai*) are created for different areas of expertise, and these subcommittees are further subdivided into

Table 4.1 Laws addressing automotive technological policy (including laws covering basic, safety, environmental pollution, noise pollution, and vibration restrictions)

Category	Legislation	Overview	Supervising authority
Basic / Safety	Road Transport Vehicle Act (Act No. 185 of June 1, 1951)	Introduces a vehicle inspection system that seeks to ensure vehicle safety, prevent pollution and other environmental damage, and improve technology for vehicle maintenance.	MLIT
	Road Traffic Act (Act No. 105 of June 25, 1960)	Seeks to mitigate hazardous road conditions and ensure a smooth and safe flow of traffic.	NPA
Air pollution	Air Pollution Control Act (Act No. 97 of June 10, 1968)	Regulates emissions of hazardous air pollutants from factories and sets emissions standards for vehicle emissions. Also stipulates that businesses may be liable for the health hazards of pollution.	MLIT, MOE
	Act concerning Special Measures for Total Emission Reduction of Nitrogen Oxides and Particulate Matter from Automobiles in Specified Areas (Act No. 70 of June 3, 1992)	Formulates policies and plans for reducing the total amount of NO_x and PM emissions from motor vehicles in specific areas. In addition, imposes special restrictions on NO_x and PM emissions from certain vehicles in certain areas.	MLIT, MOE
	Act on Regulation, etc. of Emissions from Non-road Special Motor Vehicles (Act No. 51 of May 25, 2005)	Controls emissions of exhaust gas from special (off-road) motor vehicles.	MLIT, MOE, METI

Continued

Table 4.1 Continued

Category	Legislation	Overview	Supervising authority
	Act on the Prevention of the Generation of Particulates from Studded Tires (Act No. 55 of June 27, 1990)	Restricts the use of studded tires.	MOE
	Act on Compensation, etc. of Pollution-related Health Damage (Act No. 111 of October 5, 1973)	Compensates persons suffering pollution-related health damage; introduces measures to prevent pollution-related health damage.	MOE
Noise / Vibration	Noise Regulation Act (Act No. 98 of June 10, 1968)	Sets maximum permissible limits on automotive noise emissions, including near factories and construction sites.	MLIT, MOE
	Vibration Regulation Act (Act No. 64 of June 10, 1976)	Restricts ground vibrations due to road traffic, as well as ground vibrations due to factories or construction projects.	MOE

Notes: **MLIT**: Ministry of Land, Infrastructure, Transport and Tourism; **NPA**: National Police Agency; **MOE**: Ministry of the Environment; **METI**: Ministry of Economy, Trade and Industry.
Source: Prepared by the authors based on the documents from the web sites of the relevant ministries and agencies of Japan.

sectional committees (*bunkakai*) and working groups. Each committee includes participants from all member companies and is chaired by an executive-level employee of a member company. The committees identify problems to be addressed by the industry and spearhead the development of solutions to these problems, including the process of choosing a 'base stance.' Substantive policy details are analyzed at the subcommittee and sectional committee levels. All member companies also assign employees with expertise in particular areas to serve as members of the appropriate sectional committees, and engage themselves in substantive activities, such as gathering information and making inquiries, to assist in addressing problems. The subcommittees most relevant to decisions on automotive technology policies are those established under the Safety & Environmental Technology Committee, the

Table 4.2 Laws addressing automotive technological policy (including laws covering global environmental issues, roads, and other matters)

Category	Legislation	Overview	Supervising authority
Global environment	Law concerning the Protection of the Ozone Layer through the Control of Specified Substances and Other Measures (Act No. 53 of May 20, 1988)	Introduces measures to restrict the manufacturing and use of ozone layer-depleting substances, as well as to control exhaust emissions of these substances.	MOE, METI
	Act on Ensuring the Implementation of Recovery and Destruction of Fluorocarbons concerning Designated Products (Act No. 64 of June 22, 2001)	Implements measures to provide for the secure recapture and destruction of fluorocarbons from industrial products; clarifies corporate liability.	MLIT, MOE, METI
	Act on the Rational Use of Energy (Act No. 49 of June 22, 1979)	Implements measures to advance the rationalization of the use of energy, including measures intended for factories, the transportation sector, buildings, and general machines and apparatus.	MLIT, METI
	Act on Recycling, etc. of End-of-Life Vehicles (Act No. 87 of July 12, 2002)	Mandates proper processing of, and disposal of waste from, end-of-life vehicles, including recycling of parts where possible.	MOE, METI
Road and etc.	Road Act (Act No. 180 of June 10, 1952)	Clarifies maintenance of the road network; creates a structure by which roads are planned, constructed, maintained, and funded.	MLIT

Continued

Table 4.2 Continued

Category	Legislation	Overview	Supervising authority
	Road Transportation Act (Act No. 183 of June 1, 1951)	Clarifies management of road transportation; discusses services to ensure adequacy of roads and appropriate response to the needs of road users.	MLIT
	Traffic Safety Policies Basic Act (Act No. 110 of June 1, 1970)	Specifies the traffic safety obligations of the national government, of local public entities, and of transport sector corporations. Introduces traffic safety planning, including measures undertaken by both national and local governments.	NPA
	Act concerning the Securement of Parking Space (Act No. 145 of June 1, 1962)	Requires vehicle owners to secure parking spaces. Mandates that roads are not to be used as storage for vehicles. Strengthens parking restrictions.	MLIT, NPA
	Act on Special Measures concerning Road Construction and Improvement (Act No. 7 of March 14, 1956)	Provides special measures for construction, maintenance, repair, and other management of toll roads.	MLIT

Notes: **MLIT**: Ministry of Land, Infrastructure, Transport and Tourism; **NPA**: National Police Agency; **MOE**: Ministry of the Environment; **METI**: Ministry of Economy, Trade and Industry.

Source: Prepared by the authors based on Japanese legislation.

Environment Committee, and the Traffic Affairs Committee. Figure 4.1 depicts the organizational structure of these committees.

Each subcommittee corresponds to one or more regulations addressing automotive technological policy. For instance, the regulation of automobile emissions is handled primarily by the Emissions & Fuel Efficiency Subcommittee, which is the contact point for the regulatory

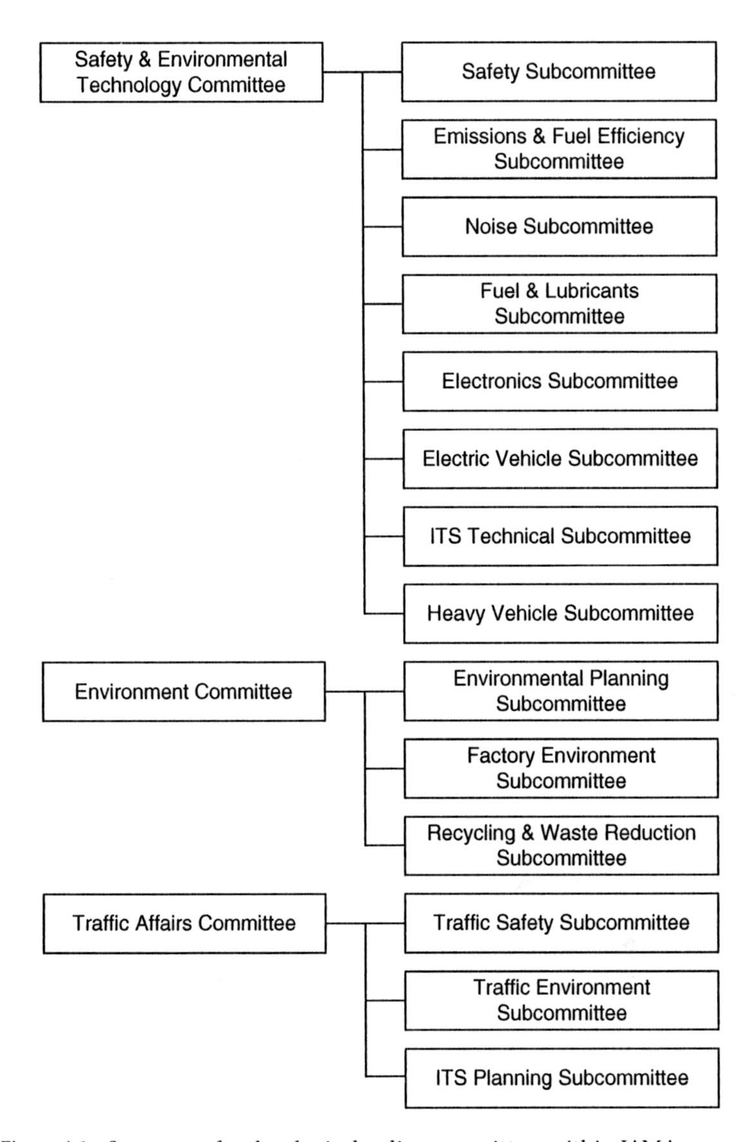

Figure 4.1 Structure of technological policy committees within JAMA

Source: Prepared by the authors based on the documents from the web site of the Japan Automobile Manufacturers Association, Inc. (JAMA), http://www.jama.or.jp/intro/organize/index.html

authority. These subcommittees also serve as points of contact for the associated councils.

Comparing the structure of these committees and subcommittees with that of the administration bureaus and councils that are in charge of automotive technology policies, we see that the two organizational structures (including the role and authority assigned to each entity) are quite similar, reflecting the system of close cooperation among industry (JAMA), government (ministries) and academia (councils). This close relationship has existed and been cultivated ever since the establishment of JAMA.

As a particular example of the collaboration among industry, government, and academia, let us now review the case of automobile emission regulations. The government first consults a council to examine the targets, standards, and timing of the regulation. The council then establishes subgroups,[2] including expert committees and working groups, and recruits technical specialists, primarily from universities, to serve in these subgroups. Research then proceeds for around two years, including the solicitation of input from industry and the preparation of a draft recommendation focusing on technical details. Decisions made by expert committees and working groups are generally adopted, without significant modification, as the recommendations of the council, and go on to be reflected in automotive emission regulations. Most members of the expert committees and working groups engage in a close exchange of information with the automobile industry. JAMA also sets up sectional committees to serve as points of contact for the expert committees and working groups under the Emissions & Fuel Efficiency Subcommittee, responding to hearings and exchanging information with expert committee members. An important aspect of this process is that all car-makers have already responded to the regulations created by the process even before those regulations come into effect. This indicates the extent to which regulatory standards are established with a clear eye to the degree of technological advancement that can be expected at the time the regulations are implemented, a feature made possible by detailed discussions at the council and expert committee levels (on topics such as, for example, prospects for the status of future technology development) and by the creation of short-term guidelines for technology development.

The content of other regulations, such as those addressing fuel economy and seat belts, is drafted in more or less similar ways (except for fuel economy requirements under the top-runner approach, discussed below). It is clear that this method of drafting regulations, with its close collaboration among industry, government, and academia, has been responsible for accurate predictions of technological advancement, and technological trends, at the time the regulations have gone into effect. We also believe that this industry–government–academia collaboration, with its fine-grained analysis of technical details, has been one of the most important elements underlying the international competitiveness of the Japanese automobile industry.

This method of drafting regulations differs significantly from that practiced in the US. In the US, regulatory standards are based entirely on the levels required to achieve the objective of the regulations, with no attention paid to the status of technology development. Instead, longer target periods are set than in Japan, allowing time for technologies to advance to levels needed to comply with regulations. In practice, the Clean Air Act (Muskie Act) of 1970, and its 1990 revision, established regulatory standards that were impossible to achieve with the technology available at the time; no manufacturer succeeded in complying with the standards by the time they were to go into effect, and the government was forced either to extend the deadline or to roll back the regulations.

4.2.2 The top-runner approach

The 'top-runner' approach to crafting technology policies differs from the processes discussed above. This procedure, which was introduced in 1999 as part of a revised Act on the Rational Use of Energy, is a technique for setting energy-saving standards for electronic products and fuel efficiency and emissions standards for automobiles at the most stringent levels existing in the market. The law also mandates that the names of companies who continue to sell products that fail to meet the standard, and the names of the non-complying products, be make public as a penalty; the companies are also fined. Passenger cars were first designated as energy-saving devices, with target fuel economy standards for 2010 set under the top-runner approach. A strengthened standard for target year 2015 is currently in place to achieve further improvements in fuel economy.

The target fuel economy standard for 2015 was formulated, and the details of its specific regulations crafted, in meetings over a period of two years by a deliberative body staffed by the two ministries that administer the Act on the Rational Use of Energy, namely, the Ministry of Economy, Trade and Industry and the Ministry of Land, Infrastructure, Transport and Tourism. This deliberative process identified the vehicle exhibiting the best fuel economy performance of all vehicles in the market, adopted its fuel economy as a reference point, and then investigated the prospects for fuel economy improvements due to technological innovations before the target year, as well as the potential negative impact of emissions regulations. The examination process conducted hearings with car-makers and other industry players regarding technologies for improving fuel economy, allowing policymakers to assess the degree of future fuel economy improvements that could be expected from technological advances. Table 4.3 summarizes the results of these assessments.

The table makes quantitative assessments of the impact of various technologies on fuel economy. However, not all technologies are applicable to all types of vehicle. Thus, when setting actual values for fuel economy standards, anticipated levels of fuel economy improvement are estimated by considering expected market penetration rates for new technologies and the extent of their utilization in top-runner vehicles.

As discussed above, in the top-runner approach, decisions on the content of regulations are made in close collaboration between government and industry. Both sides take careful steps to avoid the negative effects that top-runner technologies can have on the marketplace. In this sense, the process differs little from the procedure by which other regulations are crafted.

As for questions relating to the reasoning underlying regulatory measures and the decision-making process through which they are formulated, the *public comment* provisions mandated by the 2005 revision of the administrative procedure rules have recently made it much easier for detailed information to be released to the public, and this information will undoubtedly be used as an important resource for formulating and evaluating policy in the future.

4.2.3 The role of regulatory policy

As described above, Japanese automotive technology regulations are devised through close collaboration among the automotive industry, the Japanese government, and academic institutions, and, once formulated, all regulations have been enacted without relaxation or rollback. Japanese car-makers have ensured the development of all technologies needed to meet regulations by their dates of enactment. In other words, the content of regulations has been chosen specifically to enable car-makers to comply with them. The advent of the 'top-runner' approach has not changed this basic feature in any essential way.

In the US, on the other hand, although regulations lie at the heart of automotive technology policies just as in Japan, the method used to

Table 4.3 Impact of fuel economy technologies

Fuel economy improvement factors and potential negative impact factors			**Improvement rate (%)**
Fuel economy improvement factors	Further improvement in existing technologies for improving fuel efficiency	High-compression engines	2–4 in total
		Friction reduction	
		Weight reduction	
		Reduction in running resistance of vehicle body	
		Low-rolling-resistance tires	
		Optimization of total engine control	
	Engine improvements (Gasoline engines)	4 valves	1
		2 valves with 4-point ignition	2
		Variable valve timing mechanism	1–7
		Direct-injection stoichiometric engines	2
		Direct-injection lean-burn engines	10
		Modulated cylinders	7
		Miller cycle	10
		Massive EGR	2
		Roller cam follower	1
		Offset crank mechanism	2
		Variable compression ratio	10
	Engine improvements (Diesel engines)	4 valves	1
		Electronic fuel injection	1.5
		Common-rail	2.5
		Direct-injection diesel engines	8
		High-pressure fuel injection	1
		Turbo charger and turbo efficiency improvement	2–2.5
		Intercooler	1
		EGR	0.5–1
		Roller cam follower	1.5
		Offset crank mechanism	2

Continued

Table 4.3 Continued

Fuel economy improvement factors and potential negative impact factors			**Improvement rate (%)**
	Reduction in losses due to accessories	Electric power steering	2
		Charge control	0.5
	Power train improvements	Idling control in neutral	1
		Multi-speed AT	1–4
		CVT	7
		Automated MT, dual clutch transmission	9
		MT	9
	High-fuel-efficiency vehicles	Hybrid vehicles	15–70
		Diesel engine vehicles	20
		Idling stop vehicles	4–7
Potential negative impact factors	Emissions regulations	Degradation in fuel economy due to (1) engine improvements needed to comply with 2009 emission control regulations for diesel and direct-injection lean-burn vehicles (NO_x reduction due to EGR improvements, PM reduction due to high-pressure fuel injection, etc.), (2) NO_x storing reduction catalysts, and (3) post-treatment devices such as DPF for continuous regeneration methods	3–7.5
	Safety measures	Degradation in fuel economy due to increased weight and running resistance needed to address offset crashes, pedestrian protections, and ISO-FIX	0.1–1.4
	Noise control		0.1

Source: Prepared by the author based on Ministry of Trade, Economy and Industry (2007).

craft these regulations is different: regulations in the US are formulated with primary emphasis placed on coordination and consensus-building among stakeholders, rather than on technological feasibility. When US car-makers have failed to develop needed technologies by regulatory deadlines, either the regulations have been suspended or their deadlines have been extended.

The history of the development of automotive technology and automotive technological policy in post-war Japan demonstrates that, although the influence of the US on the introduction of safety regulations, emissions limits, fuel economy standards and other measures has been significant, the development of technology has been primarily characterized by the process of emphasizing technological feasibility and determining the content of regulations through close collaboration among industry, government, and academia. A survey of Japan's regulatory policies today reveals no serious failure in policies or the methods used to craft them, and this has been one key factor underlying the international competitiveness of Japan's current automotive technologies.

4.3 The current state of the Japanese automobile tax system – and some of its problems

In this section we survey the current state of the Japanese automobile tax system and discuss some of the problems from which it suffers.

4.3.1 The current state of the automobile tax system

a) Taxable items and their scales.

The structure of Japan's automotive tax system as of 2007, including items taxed, tax rates, earmarked destinations for tax revenue, and other features, is detailed in Table 4.4. In the current automotive tax system, taxes are imposed at each phase in a vehicle's life cycle: acquisition, ownership, and actual travel. There are eight taxable items in total, and total tax revenue in 2007 was 7,684.10 billion yen.

Revenue from six out of the eight items taxed (all but the Automobile Tax and the Light Vehicle Tax) had been earmarked for road construction by the end of fiscal year 2008; use of this revenue was restricted to road maintenance and improvement. This system of earmarking tax revenues for road construction imposed taxes based on the benefit principle; it began in 1953, with the earmarking of revenue from the gasoline tax for road construction, in an effort to promote urgent and

Table 4.4 Automotive taxes in Japan (2007)

When tax imposed	Title of tax	Type of tax	Use of tax revenue	Tax structure	Tax rate	2007 tax revenue (100 million yen)
Acquisition	Automobile Acquisition Tax	Prefectural tax	Tax revenue earmarked for road construction by prefectural and municipal governments	Tax assessed on acquisition value at the time of purchase (Vehicles whose value is 500 thousand yen or less are exempt)	Private vehicles: 5% Commercial vehicles and mini-sized vehicles: 3%	4,855
Ownership	Automobile Tonnage Tax	National tax	Two-thirds of tax revenue goes to the national government (of which 77.5% is earmarked for road construction), with the remaining one-third earmarked for road construction by municipal governments	Tax assessed on weight of vehicle at time of each automobile safety inspection	Private vehicles: 6,300 yen/year per 0.5t Commercial vehicles: 2,300 yen/year per 0.5t	10,740
	Automobile Tax	Prefectural tax	General revenue for prefectural governments	Tax assessed on vehicle owner on April 1 of every year	(Example) Private vehicles (1,001–1,500 cc): 34,500 yen/year)	17,477

Continued

Table 4.4 Continued

When tax imposed	Title of tax	Type of tax	Use of tax revenue	Tax structure	Tax rate	2007 tax revenue (100 million yen)
	Light Vehicle Tax	Municipal tax	General revenue for municipal governments	Tax assessed on vehicle owner on April 1 of every year	(Example) Private mini-sized vehicles (Four-wheel vehicle): 7,200 yen/year	1,636
Travel	Gasoline Tax	National tax	Earmarked for road construction by national government	Tax assessed on gasoline purchases	48.6 yen/l	28,449
	Local Road Tax		Earmarked for road construction by prefectural and municipal governments		5.2 yen/l	3,044
	Diesel Handling Tax	Prefectural tax	Earmarked for road construction by prefectural governments	Tax assessed on light oil purchases	32.1 yen/l	10,360
	Petroleum Gas Tax	National tax	50% of revenue earmarked for road construction by national government; remaining 50% earmarked for road construction by prefectural government	Tax assessed on LP gas purchases	17.5 yen/kg	280

Note: The 'Light Vehicle Tax' is a tax on mini-sized vehicles.

Source: Prepared by the authors, based on Japan Highway Users Conference, *Road Administration 2007*, and the web pages of the Japanese Ministry of Finance and Ministry of Internal Affairs and Communications.

systematic improvement of Stone Age roads in Japan. Since then, and particularly following the First Road Improvement Five Year Plan initiated in 1954, the government has increased tax rates and introduced new taxes in response to increasing need for road investments. During the early stages of economic development, the system worked well to ensure budget stability. However, with economic growth came criticisms of the system for its budgetary inflexibility and its inefficiency in investments. Although opinions had clashed since the 1970s as to whether funds earmarked for road construction should be incorporated into the general budget or kept separate, the issue became a matter of critical significance in the Koizumi cabinet. In May 2008, the Fukuda cabinet adopted a policy to incorporate the entire quantity of funds earmarked for road construction into the general budget in the year 2009. In this way, the system of earmarking automotive tax revenue for the maintenance and construction of roads – which had been in place continuously for 55 years since 1954 – was abandoned in April 2009.

In contrast, revenues from the Automobile Tax and the Light Vehicle Tax have been categorized as general revenue for local governments – with use not formally restricted to road construction – since the introduction of these taxes. However, in practice, prefectural and municipal governments commonly pour large portions of general revenue – more than the total revenue from both taxes – into the maintenance and improvement of roads. Thus revenues from the Automobile Tax and the Light Vehicle Tax have *effectively* been earmarked for roads, as (until recently) was explicitly the case for other automotive tax revenues.

Thus, as of October 2010, the automotive tax system in Japan remains largely as detailed in Table 4.4, with two exceptions: funds formerly earmarked for roads are now allocated for general use, and a 'green tax' system has been introduced, as discussed below.

b) The green tax system and 'eco-car' subsidies.

We here discuss the 'green tax' system, under which vehicles with smaller environmental footprint enjoy relatively privileged status. The first step in this direction was taken with the Automobile Acquisition Tax. In 1975, the rate of this tax was reduced for electric vehicles, in an effort to popularize low-emission vehicles, and rate reductions were subsequently extended to hybrid cars, natural gas vehicles, and other cleaner technologies. An additional tax deduction system designed to encourage the spread of high-fuel-efficiency vehicles has been in place since 1999.

Next, in 2001, the structure of the Automobile Tax was reorganized, in a revenue-neutral way, to increase the tax burden on vehicles with heavier environmental impact, and reduce the burden on lighter-impact vehicles. The goals of this reorganization were to reduce CO_2 emissions and combat NO_x and PM pollution, and the details of the program, such as the precise conditions a vehicle must satisfy to receive favorable treatment, have been adjusted several times since its introduction. In addition, under this system, older vehicles (defined as vehicles that were first registered as new vehicles 11 or more years ago (for diesel vehicles) or 13 or more years ago (for gasoline vehicles) are taxed more heavily.

In April 2009, a program of reduced taxes for 'eco-cars' was introduced in conjunction with this green tax system. The Automobile Tonnage Tax was added to the list of taxes for which tax reductions based on environmental friendliness were available, and the tax reductions available for the Automobile Acquisition Tax were significantly enlarged. Table 4.5 details some attributes of this 'new green tax' system, as it came to be known. Hybrid cars are granted full exemption from the Automobile Acquisition Tax and the Automobile Tonnage Tax. In addition to hybrid cars, ordinary-sized/small-sized vehicles and mini-sized vehicles that meet certain conditions can also qualify for 50–75 per cent reductions of the Automobile Acquisition Tax and the Automobile Tonnage Tax. The tax reductions for eco-cars are a temporary measure, in place only from April 2009 through fiscal year 2012.

These provisions of the 'new green tax' system dovetailed with the introduction of 'eco-car' subsidies to encourage consumers to adopt more environmentally friendly vehicles. This subsidization system, designed to encourage drivers to trade in older vehicles for more environmentally friendly 'eco-cars' and introduced to achieve both environmental and economic objectives, came into effect in June 2009. All vehicles newly registered between April 10, 2009 and September 30, 2010 are eligible for the trade-in subsidy, and drivers must commit to using the new vehicle for at least one year after registering it. Because the program was allocated a fixed budget, the subsidies will end as soon as the budget is exhausted; for fiscal years 2009 and 2010, a combined total of 583.7 billion yen was allocated for private vehicle subsidies. A driver trading in a vehicle 13 or more years old (as measured by the date of first registration) for an eco-car would receive a subsidy of 250,000 yen for a passenger car (an ordinary-sized/small-sized passenger car) or 125,000 yen for a mini-sized vehicle (K-car). A driver purchasing a new eco-car would receive a subsidy

Table 4.5 The new green tax system (gasoline-powered passenger car)

Vehicles Affected		**The amount of tax reduction**		
Conditions on exhaust gas emissions	**Conditions on fuel efficiency**	**Automobile Tax: Applies to vehicles registered before the end of March 2012 (applicable for one year)**	**Automobile Acquisition Tax (for new vehicles): Applies to vehicles registered between April 2009 and the end of March 2012**	**Automobile Tonnage Tax: Applies to vehicles registered between April 2009 and the end of April 2012**
75% or greater reduction in emissions over 2005 standard levels	Fuel efficiency at least 25% higher than 2010 standard level	50% tax reduction	75% tax reduction (100% tax reduction for hybrid cars)	75% tax reduction (100% tax reduction for hybrid cars)
	Fuel efficiency at least 20% higher than 2010 standard level	No tax reduction	50% tax reduction	50% tax reduction
	Fuel efficiency at least 15% higher than 2010 standard level	No tax reduction	50% tax reduction	50% tax reduction

Source: Prepared by the authors based on the documents from the web site of the Ministry of Land, Infrastructure and Transport (http://www.mlit.go.jp/)

of 100,000 yen for an ordinary-sized/small-sized vehicle and 50,000 yen for a mini-sized vehicle. As of August 2010, some 2.4 million ordinary-sized/small-sized vehicles and 960,000 mini-sized vehicles had been deemed eligible for the trade-in subsidies. In addition, some local (prefectural or city-level) governments have instituted their own eco-car subsidies in addition to those provided by the national government.

4.3.2 Problems in the automotive tax system

Faced with continual need to expand the pool of resources available for maintaining road facilities, the Japanese government has introduced so many revisions to Japan's automotive tax system over the years – including both the imposition of new taxes and increases in existing taxes – that the system today can only be described as something of a crazy patchwork quilt. In addition to its excessive complication, the Japanese automotive tax system suffers from a number of fundamental structural problems, as has been pointed out by many observers. In this section we will discuss these difficulties, focusing in particular on the problematic balance between fixed rates and metered rates.[3] The discussion in this section will not address the green tax system and the 'eco-car' subsidies that have recently been introduced as temporary measures.

As we explained above, the automotive tax system is designed to collect taxes based on the benefit principle. In other words, although we use the term 'tax,' in fact taxes are nothing more than 'fees' paid by automobile users for their travels.

Considered as fees that automobile users pay for road travel services, automotive taxes have a dual-tariff structure, with both basic and metered rates. The Automobile Acquisition Tax, the Automobile Tonnage Tax, the Automobile Tax, and the Light Vehicle Tax are assessed at fixed rates irrespective of miles traveled. On the other hand, the energy taxes (the Gasoline Tax, the Local Road Tax, the Diesel Handling Tax and the Petroleum Gas Tax) are assessed at metered rates, depending on the quantity of road travel services used (which can be measured by energy consumed).

When this dichotomy is applied to Table 4.4, the total revenue from taxes assessed at fixed rates amounts to 3,470.8 billion yen, or 45.1 per cent of total car-related tax revenue. Total revenue from taxes assessed at metered rates is 4,213.3 billion yen, or 54.9 per cent of total car-related tax revenue.

Miyoshi (2001) reviewed problems in this automotive tax structure from two perspectives: (a) two guiding principles for establishing levels for public service fees, namely, the cost-of-service principle and the value-of-service principle, and (b) compliance with environmental requirements. The remainder of this subsection is based on the analysis of Miyoshi (2001).

Consider first a system based on the cost-of-service principle, in which the cost required to supply a service to an individual user is borne by that user. For electric and gas utilities, for instance, each individual user is a source of fixed individual cost regardless of service usage – for example, the cost of measuring usage or installing distribution lines. Fixed rates are thus rational in these cases. In the case of road travel services, on the other hand, there is no fixed fee associated with an individual user. Next consider a system based on the value-of-service principle, in which we charge fees based on the value derived by each individual user from the service concerned. In the case of telephone service, for example, users derive utility from being able to receive calls, even if they do not make phone calls themselves. But this type of utility also has no analogue in road travel services.

Even if we can accept the existence of fixed rates, the dual-tariff system will be appropriate as long as no consumer is forced out of the market by the fixed rates. However, if the fixed rates are high, as in Japan's current automotive tax system, then small-scale users will be excluded from the market. In addition, there will be a large internal subsidy effectively paid to heavy road users by less frequent road users.

Finally, let us examine the automotive tax system from an environmental perspective. As discussed above, fixed-rate taxes account for 44.5 per cent of all car-related tax revenues. The cost per mile traveled of such a tax decreases with increasing distance traveled. But to preserve our environment we should assess fees that *increase* with increasing distance traveled. Evidently the current tax structure is the precise opposite of what would be desired from an environmental perspective.

In addition to these points, Miyoshi (2001) pointed out several other aspects of Japan's existing automotive tax system that are difficult to justify on the basis of environmental considerations *or* in terms of the basic principles governing the assessment of public utility fees. Such aspects include the tax gap between commercial and private use vehicles (at both the acquisition and ownership phases) as well as the gaps between travel-phase taxes on diesel and gasoline vehicles.

4.4 Future directions for automotive technology

Thus far in this chapter we have reviewed the process that has been used to date in Japan to craft automotive regulatory policy (Section 4.2) and surveyed Japan's hierarchy of automobile-related taxes, including some of the problems inherent in the structure of vehicle taxes (Section 4.3). These discussions set the stage for the present section, in which, after first reviewing the current status of the global warming crisis, we propose future policy directions for reducing CO_2 emissions.

4.4.1 Global warming crisis and the automotive technology roadmap

In the 2007 G8 summit at Heiligendamm, the leaders of the G8 nations agreed to give serious consideration to reducing global emissions of greenhouse gases by at least 50 per cent by 2050. The International Energy Agency (IEA) (2008) has anticipated the energy technologies that will need to be realized in order to achieve two scenarios: the ACT Map scenario, in which global emissions of CO_2 would return to their 2005 levels by 2050, and the 'BLUE Map' scenario, in which global emissions of CO_2 would be reduced by 50 per cent (over their 2005 levels) by 2050. We consider the automotive sectors of these technology forecasts. First, within the ACT Map scenario, the majority of the necessary improvements in energy consumption and CO_2 emission reductions within the transportation sector can come from enhanced energy efficiency of internal combustion engine (ICE) vehicles and a broader diffusion of hybrid cars throughout the vehicle market. Increased use of biofuels, particularly as a substitute for gasoline as a vehicle fuel, also plays an important role. On the other hand, the BLUE Map scenario requires that electric cars and fuel-cell vehicles number some 1 billion units worldwide by the year 2050; the transport sector will require more investment than any other single sector to fulfill the conditions of this scenario (IEA, 2008). In Japan, the Ministry of Economy, Trade, and Industry announced the *Cool Earth-Innovative Energy Technology Program* in March 2008 (METI, 2008). With the ultimate long-term goal of halving global greenhouse gas emissions (relative to their current levels) by 2050, this program identified a set of technologies on which Japan should focus its development initiatives, and established a sequence of milestones by which the progress of technology development could be definitively characterized over the long term. Within the transport sector, four particular technology initiatives were identified as critical: (a) fuel cells, (b) plug-in hybrid cars and electric cars, (c) intelligent transport systems (ITSs), and (d) production of transport biofuel. Among these initiatives, electric vehicles have been identified as a target of particular focus, with the immediate requirement being further extension of the cruising distance by the development of large capacity batteries with low cost; the technology imperative is to develop new batteries to replace today's lithium-ion cells, and the long-term goal is to have an electric vehicle with a range of 500 km, at a cost comparable to that of a gasoline-powered vehicle, by 2030.

4.4.2 The importance of evolving from direct regulations to indirect regulations (economic regulations)

As Section 4.4.1 makes clear, the global warming crisis poses a number of urgent problems that require large-scale and rapid responses. What types of government policies are best suited to meet these challenges? In this section we argue that policies must shift away from *direct* government regulation of fuel efficiency and instead toward *indirect* regulation (economic regulation) realized through tax and subsidy policies. The first reason for this is that, given the urgency of the global warming crisis and the comprehensiveness of the response it demands, it seems highly unlikely that the traditional Japanese system for crafting automotive technological policy – in which regulatory standards are determined, in collaboration between industry organizations and councils, in such a way that all car-makers are guaranteed to be able to comply – will be able to continue in its present form. There is no question that, in the past, the relationship between industry and government, and the industry organization–council system, have contributed in critical ways to ensuring the efficacy of Japanese industrial policy. Industry organizations aggregate the interests of individual corporations, achieve consensus within industrial sectors, play a critical role as a conduit for communication between government and industry (Okimoto, 1991), and help to mitigate the information asymmetry that exists between the government and any one individual corporation, thus enhancing the effectiveness of government policy (Yonekura, 1993). These general patterns are clearly visible in the particular case of automotive technological policy, in which JAMA has played a crucial role as an industry organization. However, as the global warming crisis grows more and more urgent, and as large gaps arise between car-makers in their technological capacity to develop new-energy vehicles, it seems likely that consensus among car-makers will be increasingly difficult to achieve under the traditional industry organization–council system for setting regulatory standards (in which standards are calibrated to the base technological level that will have been achieved by the time the standards go into effect, thus ensuring that all car-makers will be able to comply).

A second reason is that, when it comes to reducing CO_2 emissions, direct regulations – in the form of fuel efficiency standards – have several important disadvantages, and no particular advantages, relative to economic regulations. Direct regulations establish minimal values for regulated quantities, as well as certain objectives to be met by society as a whole, and attempt to enforce these through administrative means, enforced by legislation; most environmental regulations in Japan have come in the form of direct regulations. Because direct regulations are enforced by law, they have the advantages of certainty and immediacy, and are thus indispensable policy tools for preventing the types of pollution that relate directly to human life and health. However, in general, direct regulations suffer from three drawbacks: (1) they offer the polluter no incentive to reduce his environmental footprint beyond the levels mandated by the government; (2) because they are applied in uniform fashion regardless of the cost of compliance, they are not cost-effective measures, but instead tend to be expensive to implement for society as a whole; and (3) there are costs involved in monitoring compliance with standard levels. Let us focus here on the particular case of automobiles. Among the various societal ills caused by the use of automobiles, two problems in particular – namely, the emission of air pollutants and traffic accidents – have an immediate and direct impact on human life and health, and thus it is reasonable that direct government regulation has been the primary policy response to these two issues. On the other hand, when it comes to the challenges of reducing energy consumption and controlling CO_2 emissions, there is no particular reason to favor a policy approach based on direct regulations, such as fuel economy standards. Indeed, CO_2 is not an air pollutant, and instead is problematic primarily because of the sheer volume in which it is emitted. If we attempt to control CO_2 emissions with a program of mandated fuel economy standards, then the cost incurred by each car-maker to develop technologies to reduce CO_2 emissions will be passed on to consumers as an uniform increase in the price of all vehicles within each vehicle model. In Chapter 3 of this book, we illustrated that Japan's 2015 fuel efficiency standards have economic benefits for vehicle users. But this is only true for the *average* vehicle user; it is not the case that *all* users receive economic benefits. In fact, regulations actually amount to a transfer of income from infrequent users of automobiles to frequent users of automobiles. Moreover, because fuel efficiency regulations impose a uniform fuel efficiency standard on all car-makers, those manufacturers who have fallen behind in fuel efficiency technology are saddled with exorbitant catch-up costs, and in some situations may even be forced to exit the market. There is simply no social benefit realized by requiring consumers who only drive a few hundred kilometers a year to own fuel-efficient vehicles. Strict fuel economy standards preclude the existence of manufacturers who might offer inexpensive vehicles to such consumers, thus reducing the variety of consumer choice in the marketplace.

4.4.3 Fundamental reforms in the automotive tax system

If government policy is to evolve from direct regulations, such as fuel efficiency standards, toward indirect regulations, such as tax and subsidy policies, then the automotive tax system will need to be reformed in fundamental ways. As discussed above, the new green tax system and 'eco-car' subsidies have without question been extremely effective tools for encouraging consumers to choose more fuel-efficient vehicles, but these systems are little more than temporary measures erected atop the foundation of the existing automotive tax system. In the future, the structure of the automotive tax system must be reformed in a way that shifts away from fuel efficiency standards – a type of direct regulation – and instead toward measures to reduce CO_2 emissions.

As discussed above, several aspects of Japan's existing automotive tax system are difficult to justify on the basis of environmental considerations *or* in terms of the basic principles governing the assessment of public utility fees. In the future, basic (acquisition and ownership) taxes should be abolished or significantly reduced, while travel-phase taxes should be increased. In the remainder of this section, we will present the results of simulations conducted under our simulation model, the Road and Environmental Policy Assessment Model (Akune *et al.*, 2008).[4] This is a static model, established on the basis of data from the year 2000, which simulates the effects of automotive tax code revisions on consumer behaviors in acquiring, owning, and driving automobiles in Japan. The model is based on the original model of Kashima *et al.* (2003) with some additional features. The model is broader than those in previous studies in that it accounts for the shipping sector (trucks) as well as for vehicle acquisition and ownership behaviors for both trucks and passenger cars. The model defines five social sectors: (i) households, (ii) shippers, (iii) distributors, (iv) car-makers, and (v) the government. The behavior of each sector is adjusted by market

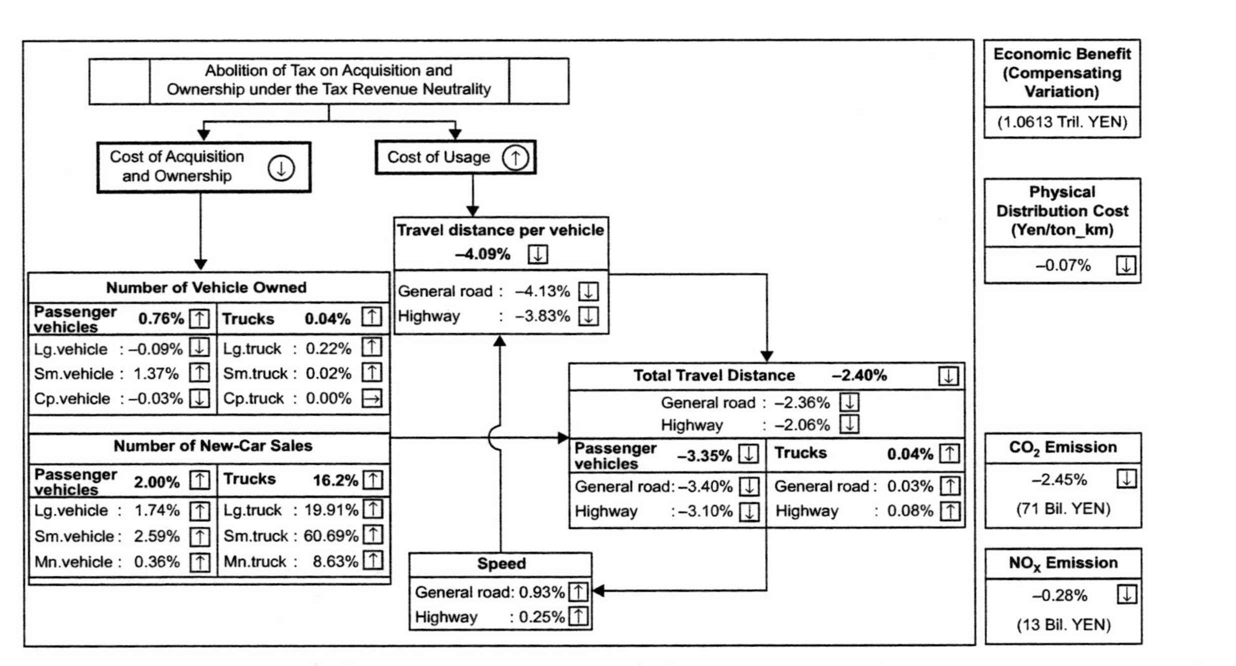

Figure 4.2 Impact of revisions to the automobile tax system

Note 1: Percentages in the figure are the percentages by which variable values in the comparison case differ from those in the base case.

Note 2: Lg., Sm., and Mn. stand for ordinary-sized, small-sized, and mini-sized, respectively.

Source: Akune et al. (2008).

mechanisms (it is a multi-sector equilibrium model), and our goal is to simulate behaviors in acquiring, owning, and driving vehicles. As our base case of simulation, we calculated sales of new vehicles, acquisition, ownership, and travel behaviors, household economic welfare, and automotive CO_2 emissions using year 2000 values for GDP, household disposable income, tax rates, and prices of goods and services. Our base case thus corresponds to theoretical results for 2000 that are reconstructed in our model. We compare these results with results predicted for an alternative scenario (the 'comparison case') in which acquisition-phase and ownership-phase taxes (the Automobile Acquisition Tax, the Automobile Tonnage Tax, the Automobile Tax, and the Light Vehicle Tax) are all abolished under tax-revenue-neutral conditions, with the amount of tax revenue thus lost made up by imposing additional taxes at the travel phase (the Gasoline Tax, the Local Road Tax, and the Diesel Handling Tax). Figure 4.2 compares the base case with the comparison case. This comparison indicates that the elimination of taxes in the acquisition and ownership phases leads to a significant increase in the number of new vehicles (the comparison case exhibits 2.00 per cent more passenger cars and 16.24 per cent more trucks than the base case), with an increase in the overall number of vehicles traded in for new vehicles. On the other hand, the increase in travel-phase taxes results in a *decrease* in miles traveled per passenger car, and reduces the total distance traveled by all automobiles (including trucks) by 2.4 per cent.[5] Consequently, CO_2 emissions decrease by 2.45 per cent, contributing to an increased economic benefit of 1.0613 trillion yen (as assessed by the method of compensating variation).

In considering the results of simulations conducted within our Road and Environmental Policy Assessment Model, including the results discussed above, it is important to keep in mind that, because this is a static model, the fuel efficiency of individual vehicles is fixed at the same value for all vehicles of a given model. The elimination of acquisition and ownership taxes and the corresponding increase in travel-phase taxes increases the attractiveness of fuel-efficient vehicles; this encourages consumers to switch to fuel-efficient vehicles, which in turn further accelerates car-makers' efforts to develop fuel-efficient technologies, and a virtuous cycle ensues. This effect can be further enhanced by dedicating a portion of the revenue derived from travel-phase taxes to subsidize the purchase of new fuel-efficient vehicles. However, a tax structure in which travel-phase taxes consist entirely of energy taxes is not necessarily optimal. This is because the advent of hybrid cars and electric cars in recent years has muddied the relationship between energy consumption and usage of roads and other travel-phase services. For such energy-saving vehicles, it will be necessary to levy fees proportional to the usage of roads and other travel-phase services. Such considerations suggest that, ultimately, the optimal tax structure will be a dual-pillared approach, including not only energy taxes but also *travel distance-based* taxes based on vehicle weight and distance traveled. The existence of an extensive infrastructure for administering the vehicle inspection system in Japan suggests that a convenient implementation of a travel distance tax might be to assess the tax at the time of vehicle inspection, based on odometer readings; but GPS-based systems should also be gradually phased in. Indeed, a GPS-based system for charging a toll for trucks based on distance traveled on *autobahns* has been in place in Germany since 2005. Moreover, in the future, if large-scale reductions in CO_2 emissions are to be achieved, it seems likely that taxes throughout the automotive tax system will have to increase. Considering the reality that tax rates, once established, cannot be easily raised, it is imperative that a host of considerations be thoroughly assessed before tax rates are fixed, including the extent to which vehicular fuel efficiency technology can be used to achieve annual emissions reduction goals, the sensitivity of travel demand to tax rates, the uses of tax revenue and the corresponding side effects, and the efficacy of various hybrid policy schemes, including alternative measures for managing traffic.

4.5 Conclusions

This chapter began with a survey of the process that has been used in the past to craft automotive regulatory policy in Japan. We noted that regulatory standard levels were generally calibrated to the base technological levels expected to have been attained by the car-makers at the time the standards went into effect, thus ensuring that no carmaker would be unable to comply with the standards or forced to exit the market. We then discussed Japan's automotive tax system, which functioned for many years as a dedicated revenue source for the construction and maintenance of roads. We pointed out that the relative magnitude of basic (mileage-independent) taxes, including acquisition and ownership taxes, is especially problematic both from an environment-friendliness perspective and on the basis of the fundamental principles governing the assessment of public utility fees. Then, based on these analyses, we considered methods for crafting automotive technological policy in response to the increasingly urgent threat posed by the global warming crisis. We reached two main conclusions. (1) If Japan continues its traditional strategy of direct government regulation of fuel efficiency, and if the process that has traditionally been used in Japan to set regulatory standards carries on into the future, then we anticipate a number of serious problems arising, including not only a failure to achieve consensus within the automobile industry but also a potential degradation in the variety of choices available to consumers in the vehicle marketplace. Thus we advocate a shift away from traditional direct regulation policies and instead toward indirect regulations (economic regulations) using tools such as tax and subsidy policy. (2) The automotive tax system must be reformed in a way that emphasizes travel-phase taxes.

In this chapter we presented the results of simulations conducted under the Road and Environmental Policy Assessment Model (Akune *et al.*, 2008). We are presently refining this model to enable quantitative predictions of the efficacy of the policies we advocate in this chapter. Our refinements consist of two main improvements to the model. First, we will add an additional category of vehicles – namely, highly fuel-efficient vehicles – to ensure that car-makers' behavior in improving fuel efficiency technology is captured as an endogenous variable. The only passenger-car categories present in the current model are ordinary-sized, small-sized, and mini-sized vehicles, and fuel efficiency is fixed at the standard levels for fiscal year 2000. Given these limitations, the model cannot realistically capture the tendency of real-world users to shift to hybrid cars and other fuel-efficient vehicles in response to policy changes such as reducing or eliminating acquisition-phase and ownership-phase taxes and increasing travel-phase taxes instead; nor can the model capture the behavior of car-makers regarding the development of fuel-efficient vehicles or improvements in fuel efficiency technology. To assess the impact of policy shifts on CO_2 emissions, we will add highly fuel-efficient vehicles to our model and introduce endogenous variables to model the fuel-efficient technology development behavior of car-makers. Second, we will refine the model to consider various different types of families within the household sector. The model used in this chapter had only one type of household, an 'average' household (the single representative household) in Japan. However, the aging of the Japanese population will have dramatic effects on household patterns of vehicle acquisition, ownership, and usage. To investigate the changes in store for the future, we are currently engaged in research that segregates households into three groups by the age of the head of household. This will enable us to understand how changes in the composition of households, as measured by the age of the head of household, affect the vehicle acquisition, ownership, and usage behaviors of the household as a whole.

Notes

1. The 14 companies are Isuzu Motors, Kawasaki Heavy Industries, Suzuki Motor, Daihatsu Motor, Toyota Motor, Nissan Motor, UD Trucks, Hino Motors, Fuji Heavy Industries, Honda Motor, MAZDA Motor, Mitsubishi Motors, Mitsubishi Fuso Truck & Bus, and Yamaha Motor.
2. Currently known as the 'Automobile Emission Expert Committee' within the Central Environment Council and the 'Automobile Emission Expert Committee' within the Air Environment Committee.
3. The fact that Japan's automotive tax system overtaxes in the acquisition and ownership phases has been observed by several authors, including Sugiyama and Imahashi (1989), Imahashi (1995) and Obuchi (1993). Imahashi (1995) showed that, by shifting emphasis to fuel taxes, the tax system could be made to approximate the pricing system, whereupon the idea of rerouting general revenues would no longer be defensible. On the other hand, to address environmental considerations, Endo *et al.* (1999) simulated the effect of reducing fuel consumption in a scenario in which 50 per cent of the ownership tax was shifted to fuel taxes under revenue-neutral conditions.
4. This model was developed by the project team composed of Yuko Akune and Hiroaki Miyoshi *et al.* in the GENDAI Advanced Studies Research Organization. The English version of Akune *et al.* (2008) can be downloaded from http://itec.doshisha-u.jp/03_publication/01_workingpaper/2007/07-31-FINAL-Akune-Miyoshi-Tanishita-itecwp.pdf (accessed November 9, 2010).
5. The model assumes a fixed volume of travel for each category of truck.

References

Akune, Y., Miyoshi, H., and Tanishita, M. (2008) 'Jidosha kanren zeisei to keizai kosei' ('Effect of Automobile Taxation System Revision in Japan'), in Miyoshi, H. and Tanishita, M. (eds), *Jidosha no gijyutu kakushin to keizai kousei: kigyo senryaku to kokyo seisaku* (*Technological Innovation in the Automotive Industry and Economic Welfare*), pp. 115–42, Tokyo: Hakuto Shobo Publishing Company.

Endo, K., Tanishita, M., and Kashima, S. (1999) 'Jidousha kanren zeisei no henkou ni yoru nenryou shouhiryou sakugen kouka no suikei shuhou no kaihatsu' (Development of Estimation Method of Effects of Reducing Fuel Consumption by the Car-related Taxation System), *Infrastructure Planning Review*, 16: 455–63 (in Japanese).

Imahashi, R. (1995) 'Douro seibi zaigen seido no kaizen ni kansuru ichi kousatsu' (Study on the Improvement of the System of Earmarked Revenues for Road Construction), *Transportation Studies*, 39: 1–10 (in Japanese).

International Energy Agency (2008) *Energy Technology Perspectives 2008: Scenarios and Strategies to 2050*, Paris: OECD Publishing.

Kashima, S., Hayes, W., Uchiyama, K., Tanishita, M., Hasuike, K., Hirota, K., Minato, K., and Miyoshi, H. (2003) *Chikyu kankyo seiki no jidousya jeisei* (*Car-*

related Taxation System in the Century of the Global Environment), Tokyo: Keiso Shobo (in Japanese).

Ministry of Economy, Trade and Industry (2007) 'Jouyousyatou no atarashii nenpikijun nikansuru saisyuu torimatome ni tsuite' (New Automobile Fuel Economy Standard for Passenger Vehicles), available at http://www.meti.go.jp/press/20070202008/nenpikijun-p.r.pdf (accessed December 1, 2010).

Ministry of Economy, Trade and Industry (2008) *Cool Earth-Innovative Energy Technology Program*, available at http://www.meti.go.jp/english/newtopics/data/pdf/031320CoolEarth.pdf, accessed November 30, 2010.

Miyoshi, H. (2001) 'Douro tokutei zaigen seido no houkousei' (Study of the Direction of the System for Earmarking Funds for Road Improvement), *International Public Policy Studies*, Osaka University, 6(1): 45–62 (in Japanese).

Obuchi, Y. (1993) *Gendai no koutsu keizaigak* (*Contemporary Economics of Transportation*), Tokyo: Chuokeizai-Sha, Inc. (in Japanese).

Okimoto, D. (1991) *Tsuusanshou to haiteku sangyou – nihon no kyousouryoku wo umu mekanizumu* (*Between MITI and the Market: Japanese Industrial Policy for High Technology*), Tokyo: Simul Publishing Company (in Japanese).

Parry, I. W. H. and Small, K. A. (2005) 'Does Britain or the United States Have the Right Gasoline Tax?', *American Economic Review*, 95(4): 1276–89.

Sugiyama, T. and Imahashi, R. (1989) 'Douro' ('Roads'), in Okuno, M., Shinohara, S., and Kanemoto, Y. (eds), *Koutsu seisaku no keizaigaku* (*Economics of Transport Policy*), pp. 207–24, Tokyo: Nikkei Publishing Inc. (in Japanese).

Yonekura, S. (1993) 'Gyoukai dantai no kinou' (The Functions of Industrial Associations), in Okazaki, T. and Okuno, M. (eds), *Gendai nihon keizai shisutemu no genryuu* (*The Japanese Economic System and its Historical Origins*), chapter 6, Tokyo: Nikkei Publishing Inc. (in Japanese).

5

Economics of Intelligent Transport Systems: Crafting Government Policy to Achieve Optimal Market Penetration

Hiroaki Miyoshi and Masanobu Kii

5.1 Introduction

In recent years, Intelligent Transport Systems (ITSs) have attracted increasing attention as revolutionary technologies for addressing the problems associated with mass automobile consumption, including traffic accidents, traffic congestion, and environmental pollution.

The term 'ITSs' is a blanket designation referring collectively to any of several transportation systems aimed at solving traffic problems by taking advantage of advanced information and telecommunications technologies.

Goods and services in the IT industry, including the Internet, cell phones, computer software, CDs, DVDs, and video games, have a common property known as *network externality*. The term refers to an economy of scale on the demand side that allows the benefit enjoyed by an individual from a good or service to increase with the number of individuals consuming that good or service. This was referred to as the 'bandwagon effect' in Libenstein (1950), as 'independent demand for a communications service' in Rohlfs (1974), and as 'direct externalities' in Katz and Shapiro (1985, 1994).

When we consider ITSs from the perspective of externality, we must pay attention to the following two properties, which are not shared by other goods and services in the IT industry. First, the traffic congestion or traffic accidents that we seek to prevent through the use of ITSs are often phenomena caused by more than one vehicle. Therefore, the existence of ITS users may reduce the risk of traffic congestion or accidents for non-users. In other words, ITSs tend to serve as a *public good*. Second, in situations where an ITS enjoys such properties, the value of the system to its users will be correctly understood only by deducting its value as a public good. This leads to interdependence between users, which differs from network externality.

In this chapter, we begin in Section 5.2 by examining certain features of the benefits of ITS technologies, taking the Vehicle Information and Communication System (VICS) as an example. In Section 5.3 we examine some of the corresponding features of the benefits of safety-related ITS technologies. Then, in Section 5.4, we explain the notion of the *optimal penetration rate* and discuss the policies required to achieve this optimal penetration rate.

Although Europe and the US are also working toward the development and practical utilization of ITS technologies, in this chapter we focus on Japanese systems. Unless otherwise noted, the analyses in this chapter are qualitative, and the figures are schematic.

5.2 Benefits and externalities of VICS

In this section we examine the externalities of the VICS.

5.2.1 Outline of VICS

VICS is a system that provides traffic-related information, such as real-time reports on traffic congestion, traffic accidents, or link travel time, to drivers through an onboard car navigation system.[1] The information is displayed on a map in the car's navigation system. This system is managed by the VICS Center, which was jointly established in 1995 by the Japanese National Police Agency, the Ministry of Posts and Telecommunications, the Ministry of Construction, and private firms. The service first started in metropolitan areas in 1996. Its service area was gradually expanded to other areas, and since February 2003 it has been available throughout Japan.

Traffic-related information, which is edited and processed at the VICS Center, is provided to drivers in one of three ways: (1) transmission through radio beacon devices installed on highways, (2) transmission through optical beacon devices installed on major open roads, and (3) FM multiplex broadcasting by FM stations in each region. Information on traffic congestion and travel time on open roads is provided through optical beacons and FM multiplex broadcasting. While FM broadcasting can provide wide-area information throughout the entire prefectural area in which the broadcasting FM station is located, optical beacons can provide high-precision information as far as 30 km ahead of, and 1 km behind, a vehicle.

In order to receive VICS traffic information, users need only install a VICS-compliant car navigation system and pay a fee of 315 yen (including a consumption tax of 15 yen) at the time of purchase. There are no additional fees or service charges. Two types of VICS unit are currently available on the open market: a 'one-media receiver,' which can receive only the FM multiplex broadcast signal, and a 'three-media receiver,' which can receive information from radio and optical beacons as well as the FM signal.

Traffic information for VICS is collected by vehicle-detector devices installed on roads. An important feature of the system is that the optical beacons, in addition to transmitting information to VICS units, also double as vehicle detectors. This feature enables optical beacons to identify individual vehicles by receiving ID-number data from vehicles equipped with the three-media receiver, whereas the other types of vehicle detectors – such as ultrasonic and microwave detectors – are limited to cross-sectional scanning to detect vehicles and their speeds. VICS thus enables optical beacons to provide higher-precision information on link travel time, as each beacon can calculate the time needed for a vehicle to travel from the previously passed optical beacon to itself.

We now consider the benefits to society resulting from the use of VICS.

5.2.2 Benefits and their beneficiaries

We divide the benefits of VICS into two classes: benefits to drivers and benefits to the environment.

The most important benefit for drivers is reduced travel time, which is realized by avoiding traffic congestion both temporally and spatially. A further benefit is reduced energy usage, which leads to reduced travel cost.[2] These benefits are enjoyed not only by users but also by non-users, as the traffic congestion suffered by non-users is alleviated when users revise their travel times and routes.

In addition to the benefits to drivers, there are two types of benefits to the environment: reduced emission of air-polluting substances, such as NO_X and PM, and reduced emission of carbon. The former decreases health damage to populations residing along major roads and reduces acid rain, while the latter helps prevent global warming.

5.2.3 Travel time reduction versus penetration rate

What is the relationship between the VICS penetration rate (the fraction of vehicles equipped with VICS units) and the travel time reduction rate? Many studies have investigated the relationship between the rate of penetration of traffic information and the travel time reduction rate for both users and non-users of the information (see, for example, Ben-Akiva *et al.*, 1991; Emmerink *et al.*, 1995; Huang and Li, 2007; Koutsopoulos and Lotan, 1990; Lo *et al.*, 1995; Mahmassani and Jayakrishnan, 1991; Yang, 1999).[3] Here, following Emmerink *et al.* (1994), we consider two types of relationship between the penetration rate of traffic information and the travel time reduction rate, as depicted in Figures 5.1 and 5.2. Both of these figures are drawn under two assumptions. The first assumption is that both information users and non-users will be advantaged by the provision of information, although some researchers have pointed out that a high level of market penetration leads to a deterioration of network-wide performance (Arnott *et al.*, 1991; Ben-Akiva *et al.*, 1991; Emmerink *et al.*, 1995; Mahmassani and Jayakrishnan, 1991). The second assumption is that Wardrop's (1952) user equilibrium is attained at a penetration rate of 100 per cent. In addition, hereafter, the values of α and β are different for different figures.

In both Figures 5.1 and 5.2, the travel time reduction rate for non-users improves with increasing penetration of traffic information, attaining 100β per cent at a penetration rate of 100 per cent. This is because the traffic will be temporally and spatially dispersed as the number of information users increases. Furthermore, traffic congestion on routes and during hours that non-users travel will be gradually relieved. For information users, Figure 5.1 shows that the travel time reduction rate for users is 100α per cent at a penetration rate of zero per cent. But this rate *decreases* with increasing penetration rate, attaining 100β per cent at a penetration rate of 100 per cent. This situation will occur when traffic information is collected by vehicle detectors installed on roads. Let us examine a case in which there are two routes, A and B, from a departure point to a destination, with Route A being more congested than Route B. In such a situation, information users will choose Route B based on traffic information. However, the number of vehicles choosing this route will grow as the number of information users increases, and, as a result, Route B will eventually become congested. The travel time reduction rate for information users will be at its greatest when the number of other users is zero, and will decrease with increasing numbers of information users. If we consider the case of VICS and assume that most information users use the one-media receiver, which can only receive the FM multiplex broadcasting information, the relationship between the penetration rate and the travel time reduction rate will be similar to that depicted in Figure 5.1.

On the other hand, in the case shown in Figure 5.2, the slope of the travel time reduction rate curve for information users changes from positive to negative at p'. This will occur when the traffic information

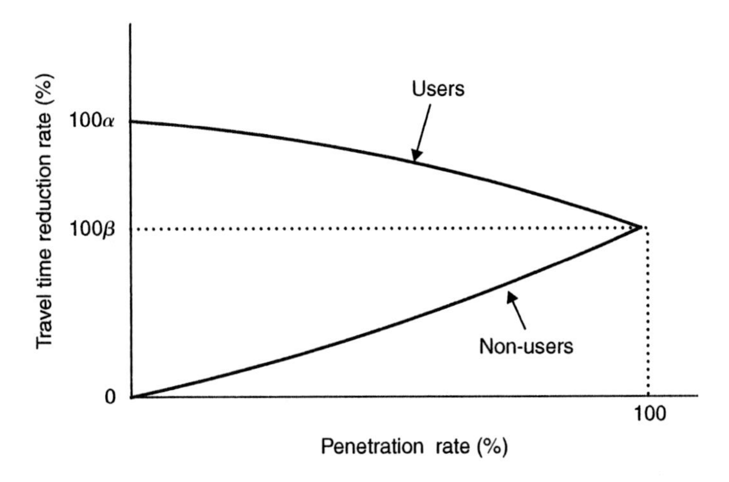

Figure 5.1 Travel time reduction rate versus penetration rate for traffic information (1)

Source: Prepared by authors based on Emmerink *et al.* (1994).

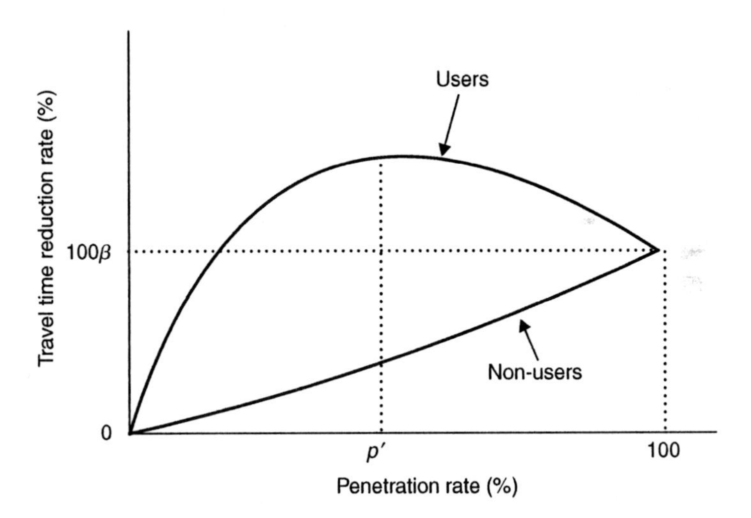

Figure 5.2 Travel time reduction rate versus penetration rate for traffic information (2)

Source: Prepared by authors based on Emerink *et al.* (1994).

is collected by onboard units. This is because, as long as the penetration rate for onboard units remains low, more information is collected from these units as the penetration rate of the devices increases, and the traffic information provided becomes more precise. However, when the penetration rate exceeds a certain level (p'), the marginal effect causing traffic congestion on alternative routes will become more important than the marginal effect improving the precision of traffic information. In such cases we can expect that the travel time reduction rate for information users will decrease with increasing penetration rate.

If we consider VICS in the case in which most information users use the three-media receiver, which can receive information from radio and optical beacons in addition to the FM broadcast signal, the relationship between the VICS penetration rate and the travel time reduction rate will be similar to that depicted in Figure 5.3, a composite of the curves in Figures 5.1 and Figure 5.2.

5.2.4 Classification of benefits

We now attempt to classify the benefits of travel time reduction, as realized with the VICS, based on a systematic attribution of the relevant benefits in each case. In order to simplify the discussion, we assume that the value of time is identical for all drivers.

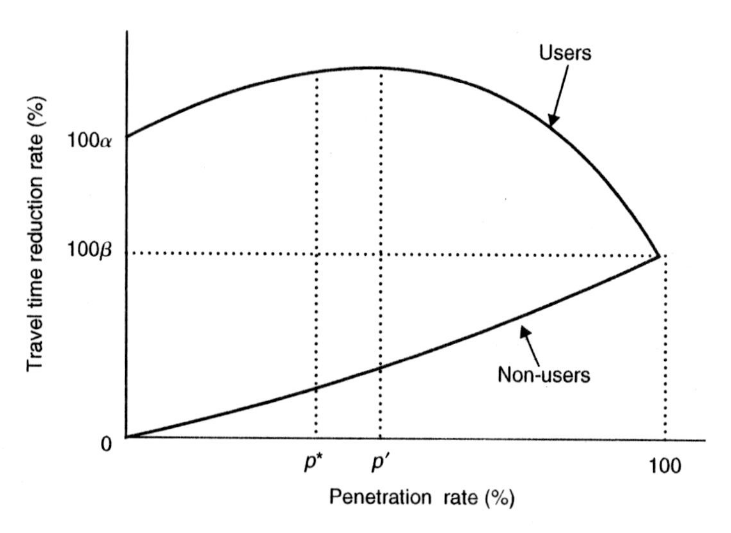

Figure 5.3 Travel time reduction rate versus penetration rate for traffic information (3)

First, the two bold-line curves in Figure 5.4 show the travel time savings benefit for one user (one vehicle equipped with a VICS unit) and for one non-user (one vehicle without a VICS unit), both as compared with a penetration rate of 0 per cent, assuming that all users use only the one-media receiver. To simplify the discussion, we assume that total travel time and the value of time are the same for all drivers. The travel time reduction rate for users decreases with an increase in the penetration rate. Thus, the overall travel time savings benefit for all users gradually decreases. On the other hand, as the travel time reduction rate for non-users increases with increasing penetration rate, the travel time savings benefit for every non-user will gradually increase.

The overall benefit of travel time reduction to society as a whole, at a penetration rate *p* of VICS, will be analyzed into three contributions: (i) the benefit for users as evaluated by the market, (ii) the benefit for users that is *not* reflected in the market, and (iii) the benefit for non-users.

First, the benefit for users as evaluated by the market is the product of two factors: (1) the margin between the travel time savings benefit for one user and that for one non-user, and (2) the number of users (which in turn is given by the product of the penetration rate and the total number of vehicles in use). This is because the market will assess the value of a VICS unit by considering the travel time savings

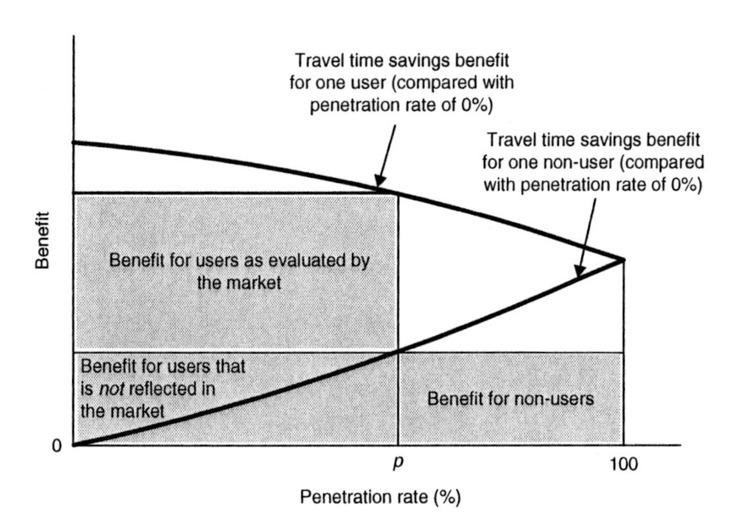

Figure 5.4 Travel time savings benefit versus VICS penetration rate (assuming most users use only FM broadcast information)

for users *as compared with the travel time savings for non-users,* not simply the absolute savings for users. Assuming a relationship between the VICS penetration rate and the travel time reduction rate similar to that plotted in Figure 5.1, the benefit for one user as evaluated by the market attains its maximum at a penetration rate of 0, and then decreases monotonically and reaches 0 at a penetration rate of 100 per cent. In other words, in this case, VICS is a system with negative network externality.

On the other hand, if most users use the three-media receiver, then the relationship between VICS penetration rate and travel time reduction rate will more closely resemble that plotted in Figure 5.3. In this case the benefit for one user as evaluated by the market now *increases* with increasing penetration, up to a boundary penetration rate of p^* ($< p'$), at which point the slopes of the travel time reduction rate curves for users and non-users become equal. Thereafter this benefit decreases monotonically, reaching 0 at a penetration rate of 100 per cent. In other words, in this case, VICS is a system whose network externality changes sign, from positive to negative, as the penetration rate increases.

In addition to the benefit for one user as evaluated by the market, users also enjoy the same travel time savings benefit as do non-users. This benefit corresponds to the part of Figure 5.4 labeled 'Benefit for users that is *not* reflected in the market.' This quantity is equal to the product of the travel time savings benefit for one non-user (as compared with a penetration rate of 0 per cent), and the number of users. This benefit increases monotonically with increasing penetration rate.

Finally, the benefit for non-users is equal to the product of the travel time savings benefit for one non-user (compared with a penetration rate of 0 per cent) and the number of non-users.

5.3 Benefits and externalities of safety-related ITS technologies

In what follows we will consider two examples of safety-related ITS technologies: the *Roadside Information-Based Driving Support Systems* (hereafter referred to as car-to-infrastructure systems), which are based on road-to-vehicle communications, and the *Inter-Vehicle Communication Type Driving Support Systems* (hereafter referred to as car-to-car systems), based on inter-vehicle communications.

Reviews of the safety and other effects of a number of ITS technologies have been produced, among others, by Kulmala (2010), Vaa *et al.* (2007), and Spyropoulou *et al.* (2008). However, to our knowledge, the present chapter is the first study to consider the effects of safety-related ITS technologies from the viewpoint of externality.

5.3.1 Changes in accident frequency vs. penetration rate

Here, we examine the frequency of accidents experienced by users and non-users of our two examples of safety-related ITS technologies, and we analyze the dependence of this frequency on the penetration rates of both systems.

a) Car-to-infrastructure systems.

The car-to-infrastructure systems are based on an infrastructure in which information collected from vehicle detectors on roads is delivered, in the form of simplified images and sounds, to the VICS unit via optical beacons. As described in Chapter 2, an improved communication protocol, with enhanced data transmission rate and the ability to communicate over longer distances, is currently planned for the future.

We examine this system in the case of collisions between vehicles making a left turn at an intersection and vehicles passing through the intersection.[4] We consider two cases: (a) the 'dual-info' case, in which both vehicles receive warnings and other information from the roadside infrastructure to help prevent accidents, and (b) the 'single-info' case, in which only the left-turning vehicle receives the warnings and other information. The systems currently under consideration in Japan are generally designed in such a way as to deliver warnings to only *one* of the two parties in a potential accident – namely, whichever party is potentially guilty of a more serious infraction, or is likely to sustain more serious damage. However, in cases in which collisions between a two-wheeled vehicle and a four-wheeled vehicle are possible, it is generally recognized that the two-wheeled vehicle should receive warnings even if it bears the lesser responsibility of the two parties, as this will help to prevent serious accidents. In addition, some car-makers are developing systems in which both parties in a potential collision always receive warnings, even if both parties are four-wheeled vehicles. In this chapter we consider both the dual-info and single-info cases.

Figure 5.5 shows the expected reduction in the number of accidents encountered versus the rate of penetration of the car-to-infrastructure system, for both users and non-users, in the dual-info case under the assumption that all drivers are homogeneous. The vertical axis in Figure 5.5 is the rate of change in the number of accidents encountered, as compared with a baseline scenario in which the penetration rate is 0 per cent.

First, for users, both types of collisions (collisions with passing-through vehicles when turning left and collisions with left-turning vehicles when passing through) will be largely avoided, irrespective of the number of system users. This is because the information needed to avoid the accident in this case is provided by devices built into the road infrastructure. The number of accidents encountered by system users will decrease by $100(\alpha + \beta)$ per cent, where, as above, β describes the reduced rate at which a vehicle making a left turn at the intersection collides with a vehicle passing through the intersection, while α describes the reduced rate at which a vehicle passing through the intersection collides with a vehicle making a left turn at the intersection. α will be equal to β under the assumption that all drivers are homogeneous. On the other hand, for non-users, the number of accidents encountered decreases as the number of users increases, simply because, when users take steps to avoid accidents, non-users avoid those same accidents. At a penetration rate of 100 per cent, the number of accidents encountered by a non-user will have decreased by $100\ (\alpha + \beta)$ per cent, which is the same reduction rate as for users.

Next, Figure 5.6 shows the expected reduction in accident frequency versus level of penetration of the car-to-infrastructure system, for both users and non-users, in the single-info case.

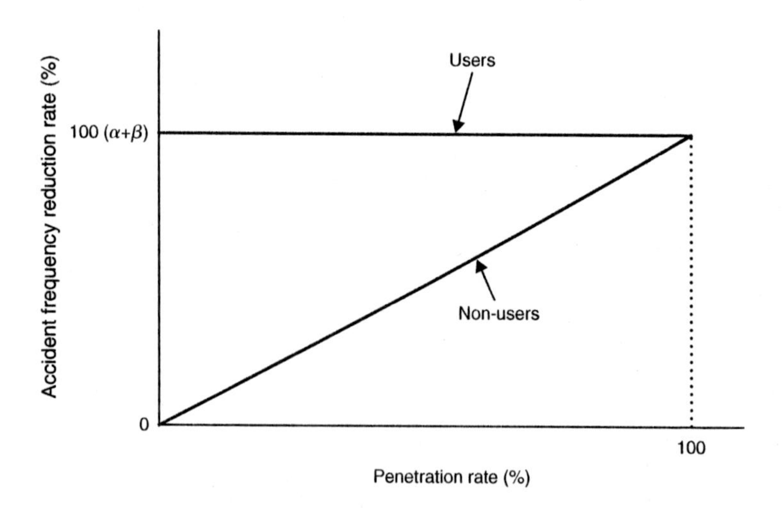

Figure 5.5 Reduction in accident frequency versus penetration rate for the car-to-infrastructure system, in the dual-info case

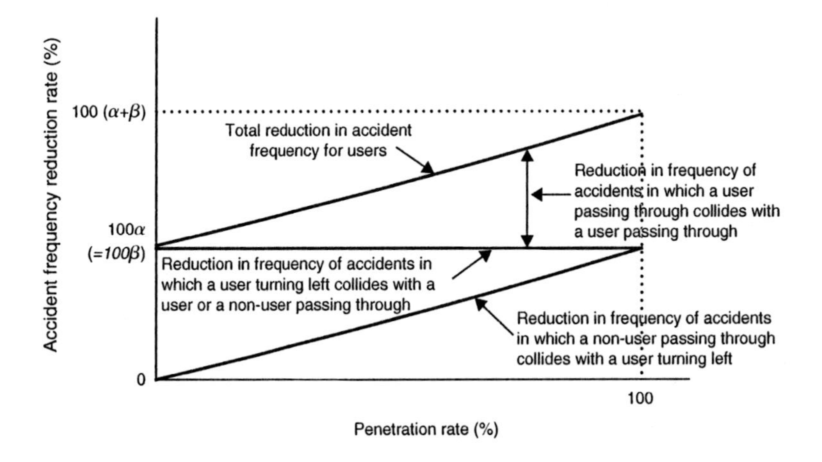

Figure 5.6 Reduction in accident frequency versus penetration rate for the car-to-infrastructure system, in the single-info case

First, for users turning left at the intersection, collisions with vehicles passing through the intersection will be largely avoided, irrespective of the number of system users. The number of such accidents encountered by left-turning users will decrease by α per cent. On the other hand, for users passing through the intersection, collisions with left-turning vehicles will decrease as the number of users increases. At a penetration rate of 100 per cent, the number of such accidents encountered by passing-through users will decrease by 100 (α + β) per cent. Meanwhile, for non-users passing through the intersection, collisions with left-turning vehicles will decrease as the number of system users increases. At a penetration rate of 100 per cent, the number of such accidents experienced by non-users passing through the intersection will decrease by 100β per cent.

b) Car-to-car systems.

The car-to-car systems are designed to avoid accidents through inter-vehicle communications. A virtue of this system is that it can be used everywhere, as long as vehicles are equipped with the communication device; there is no need to install a roadside infrastructure.

As in our previous analysis of the car-to-infrastructure case, we consider the case of collisions between vehicles making a left turn at an intersection and vehicles passing straight through the intersection. We assume that a user of the system, when making a left turn, can perceive the existence of vehicles passing through the intersection in the oncoming lane.[5] Figure 5.7 shows the relationship between the system penetration rate and the expected number of accidents that a system user will encounter, under the assumption that all drivers are homogeneous. We first consider results for system users. As the number of system users increases, the number of vehicles with which any one user can exchange information increases, and accidents for users decrease correspondingly. The number of accidents encountered decreases by $100(\alpha + \beta)$ per cent at a penetration rate of 100 per cent, where 1) α describes the reduced rate at which a vehicle passing through the intersection collides with a vehicle making a left turn at the intersection, while 2) β describes the reduced rate at which a vehicle making a left turn at the intersection collides with a vehicle passing through the intersection.

On the other hand, for non-users, the existence of system users has no significance unless we consider the unlikely event of a non-user getting caught in the middle of an accident between two other vehicles. This is because non-users cannot avoid accidents if their own vehicles are not equipped with the system.

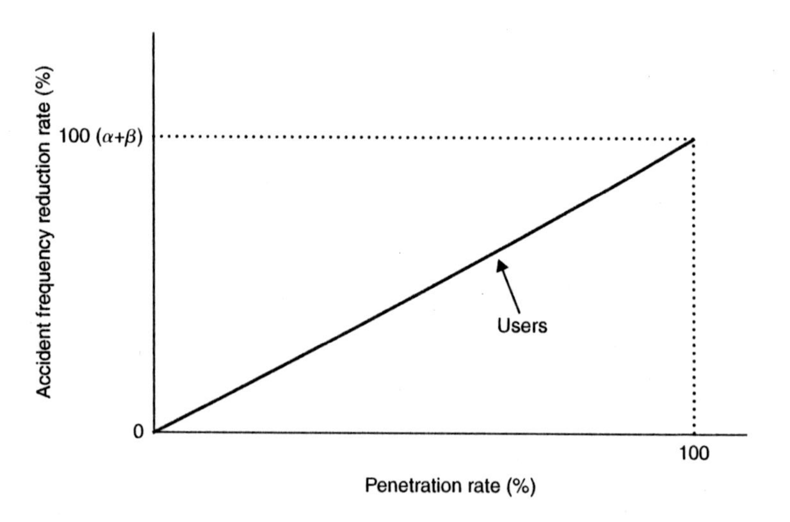

Figure 5.7 Reduction in accident frequency versus penetration rate for the car-to-car system

5.3.2 Classification of benefits

Let us now apply the theoretical concepts discussed in Section 5.2 to classify the benefits of traffic accident prevention. In order to simplify the discussion, we assume that the value of preventing traffic injuries and fatalities is identical for all drivers.

We first consider the thick curve in Figure 5.8, which plots, in schematic fashion, the benefit resulting from the prevention of traffic accidents, as realized by a user of dual info car-to-infrastructure system, versus the system penetration rate (the benefit plotted is that compared with a system penetration rate of 0 per cent).

For system users, the benefit resulting from traffic accident prevention is a constant, independent of system penetration rate, because the number of accidents encountered by any given system user does not depend on the number of system users. Meanwhile, the benefit for non-users increases in proportion to the penetration rate, because the number of accidents encountered by non-users decreases as the number of system users increases.

Next, following the theoretical discussion of Section 5.2, we will analyze the benefit of traffic accident prevention for society as a whole, at a penetration rate of p, into three contributions: (i) the benefit for users

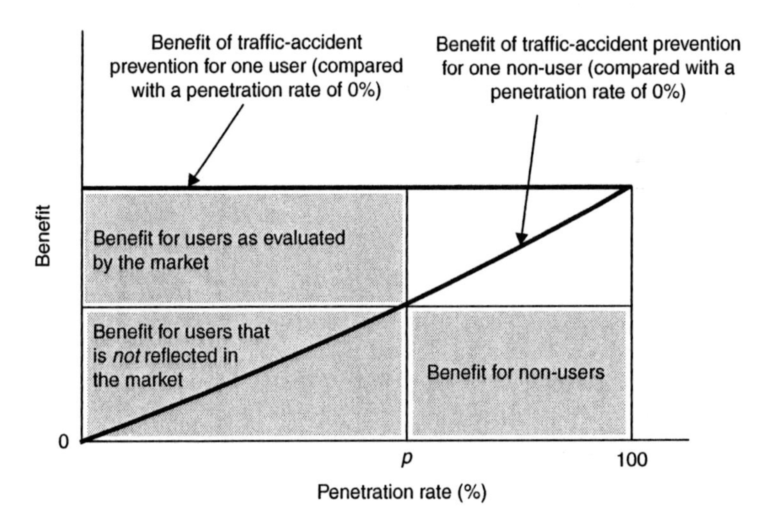

Figure 5.8 Classification of benefits of the car-to-infrastructure system, in the dual-info case

as evaluated by the market, (ii) the benefit for users that is *not* reflected in the market, and (iii) the benefit for non-users.

The benefit for users is given by the product of two factors: the traffic accident prevention benefit for one user (compared with a penetration rate of 0 per cent), and the number of users. This benefit can be further subdivided into two distinct components: the benefit as evaluated by the market and that portion of the benefit which is *not* reflected in the market.

First, the benefit for users as evaluated by the market may be thought of as the product of two factors: the margin between the benefit of traffic accident prevention for users and that for non-users, and the number of users. This is because the market will assess the value of the system by the reduction in accident frequency experienced by a user *as compared with the reduction in accident frequency experienced by a non-user,* not simply the reduction in frequency for a user in absolute terms. Assuming a relationship between system penetration rate and accident reduction rate for users as plotted in Figure 5.5, the benefit for one user as evaluated by the market attains its maximum level at a penetration rate of 0 per cent. For higher penetration rates this benefit then decreases monotonically and falls to 0 at a penetration rate of 100 per cent. In other words, in the dual-info case, the car-to-infrastructure system exhibits negative network externality.

In addition to this market-assessed portion of the benefit, users also enjoy the same benefit as non-users from the prevention of traffic accidents. This portion of the benefit for users is *not* reflected in the market, and may again be thought of as the product of two factors: the benefit of traffic accident prevention for one non-user (as compared with a system penetration rate of 0 per cent) and the number of users. This benefit increases monotonically with increasing penetration rate.

Finally, the benefit for non-users is also equal to the product of two factors: the benefit of traffic accident prevention for one non-user (as compared with a system penetration rate of 0 per cent) and the number of non-users.

On the other hand, Figure 5.9 depicts the corresponding situation for the car-to-car system. We first consider the thick curve in Figure 5.9, which plots, in schematic fashion, the benefit resulting from the prevention of traffic accidents, as realized by a user of the car-to-car system, versus the system penetration rate (the benefit plotted is that compared with a system penetration rate of 0 per cent). The traffic accident prevention benefit for each user increases gradually as the penetration rate increases, because the number of accidents experienced by any given

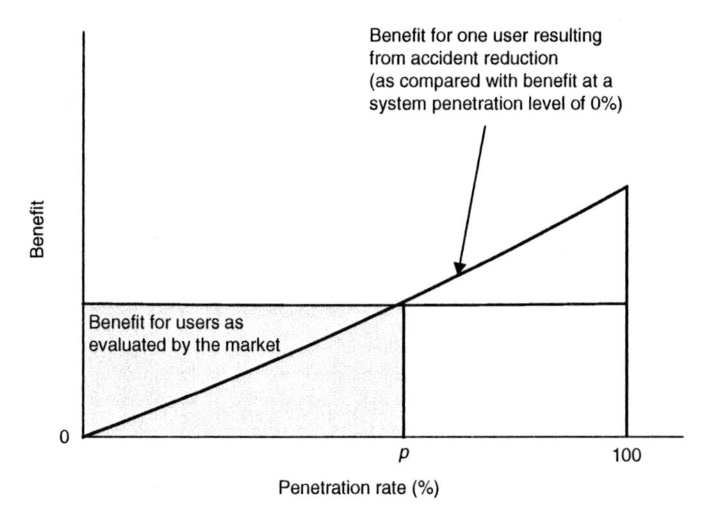

Figure 5.9 Classification of benefits of the car-to-car system

user decreases as the number of system users increases. In other words, the car-to-car system exhibits positive network externality.

The overall benefit of traffic accident prevention to society as a whole, at a penetration rate of p, is indicated by the shaded region. This is the benefit that will be evaluated by the market.

5.4 Optimal penetration rate and policies for achieving it

In this section, we will discuss the notion of the *optimal penetration rate* and explain the policies required to achieve it, for systems with positive network externality, such as the car-to-car systems. For analyses of systems with negative network externality, we refer the reader to the analyses of VICS in Chapter 6.

5.4.1 Tax and subsidy policies

Figure 5.10 shows an example of the expected equilibrium demand curve (*ED*), the curve of marginal benefit value for all users (*MUB*), and the curve of marginal social benefit (*MSB*), for a system with positive network externality, such as the car-to-car systems.

The analyses of the previous sections assumed that all drivers are homogeneous. In this section we modify this assumption as follows: the perceived benefit of the system now differs from one driver to another,

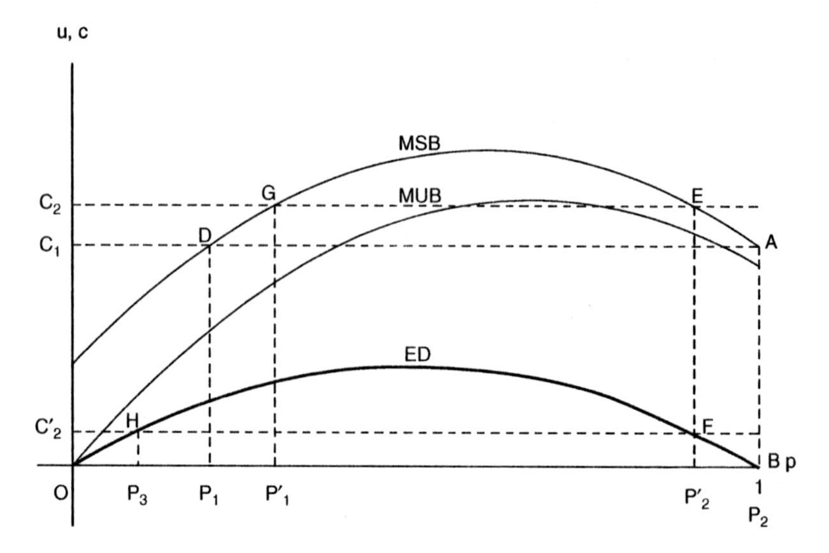

Figure 5.10 Optimal penetration rate for a system with positive network externality

and is uniformly distributed between zero and a certain positive value. We also assume that the benefit derived by a user from the system is directly proportional to the penetration rate. These assumptions are introduced following Rohlfs (1974). Assuming that drivers who perceive higher benefits will begin to use the system earlier, the perceived benefits will decrease with increasing penetration of the system. Assuming a uniform distribution of benefits and the proportionality of benefits to the penetration rate, the benefit U perceived by a marginal user at penetration rate p is given by the expression

$$U(p) = \alpha p (1 - p), \tag{5.1}$$

where p is the penetration rate and α is a proportionality constant. The expected equilibrium demand curve ED indicates the value of the benefit as evaluated by the market for a marginal user, under the condition that the expected penetration rate is equal to the actual penetration rate, namely $ED = U(p)$. The expected equilibrium demand curve passes through the origin, because the benefit for all users is zero at a penetration rate of zero, assuming that the benefit derived by a user from use of the system is directly proportional to penetration rate. ED also passes through the point (1, 0), because of the assumption that the benefit for users at penetration rate p will be uniformly distributed between zero and a certain positive value αp. In this case, the slope of *ED* is positive when the penetration rate is below a certain level. This is because the increase in benefit due to increasing penetration rate exceeds the decrease in benefit for a marginal user due to increasing penetration rate. On the other hand, the slope of *ED* changes from positive to negative at a penetration rate of 50% when the latter, negative, effect overwhelms the former, positive, effect.

The curve of marginal value of the benefit for all users, *MUB*, indicates the increase in the value of the benefit enjoyed by all users, as evaluated by the market, when the number of vehicles increases by one. When the number of vehicles increases by one, not only will this marginal user enjoy a level of benefit equal to the height of *ED* at the given penetration rate, but vehicles already equipped with the system will also be better off, because this system exhibits positive network externality. Using Eq. (5.1), *MUB* is given by the expression

$$MUB(p) = \frac{\partial}{\partial p}\left\{\alpha p \int_0^p (1-q)dq\right\}. \tag{5.2}$$

The difference between the values of *MUB* and *ED* indicates the magnitude of the benefit enjoyed by vehicles already equipped with the system.

The marginal social benefit curve, *MSB*, indicates the marginal value of the benefit enjoyed by all vehicles when the number of vehicles increases by one. *MSB* is the sum of *MUB* and the marginal value of the *other benefits* (the sum of the benefit for non-users and the benefit for users that is not reflected in the market).

In Section 5.3 we did not consider the *other benefits* of the car-to-car system, because the benefit of this system for non-users is zero. But the *other benefits* for this system will in fact be non-zero, because a decrease in travel accidents will relieve traffic congestion and thus generate travel time savings. In view of this benefit, we have drawn Figure 5.9 under the assumption that the marginal value of the *other benefits* gradually decreases as the penetration rate increases.

Finally, we assume that system usage fees are constant, independent of penetration rate, under the assumption that the production function of the system displays constant returns to scale.

Let us now examine the optimal penetration rate under these assumptions. The optimal penetration rate is the point of intersection of the price curve C (usage fee) and the marginal social benefit curve *MSB*.

Referring to Figure 5.10, first suppose the price is set equal to C_1. In this case, *MSB* intersects the price curve at the two points *A* and *D*, and the penetration rate P_2 corresponding to A, that is, 100 per cent, is the penetration rate that maximizes the social benefit. However, this penetration rate P_2 cannot be realized through market mechanisms alone. This is because the price curve lies above *ED*, so no drivers will want to use this system. In order to realize the optimal penetration rate, we will have to reduce the price of the system artificially. In other words, we will have to provide a Pigouvian subsidy equal to *AB* (=C_1) – in other words, to set the price for a user to zero. More realistically, regulatory structures, combined with subsidies to car-makers or to users, will be necessary to realize the optimal penetration rate. The same will be true for all prices less than C_1.

On the other hand, for prices above C_1, the optimal penetration rate will be below 100 per cent. Suppose the price is set equal to C_2. In this case, *MSB* intersects the price curve at the two points *E* and *G*, and the penetration rate P_2' corresponding to *E* is the penetration rate that maximizes social benefit. To realize this optimal penetration rate, we must provide a Pigouvian subsidy equal to *EF*. The net social benefit will decrease as the price increases and will fall below zero at a certain price level. If the net social benefit is negative, the optimal penetration rate is zero and there is no need to provide subsidies.

5.4.2 Creating critical mass and shifting drivers' expectations for the penetration rate

Next suppose the price is equal to C_2. By providing a Pigouvian subsidy equal to *EF*, we bring the price for users down to C_2'. In this case, the expected equilibrium demand curve *ED* intersects the price curve C_2' at the two points H and F.

H corresponds to a penetration rate known as the 'critical mass:' a dynamically unstable equilibrium point. If the penetration rate rises slightly above P_3, the market penetration rate will automatically converge to P_2', a desirable stable equilibrium point. In contrast, if the penetration rate falls below P_3, the market penetration rate will automatically converge to zero (Rohlfs, 1974). The challenge for the government and for car-makers is to achieve a penetration rate of P_3. The idea that comes immediately to mind is to distribute systems for free at a level slightly exceeding the critical mass. Another possibility is for the government artificially to increase drivers' expectations for the penetration rate by publicly announcing the target system penetration rate. The expected equilibrium demand curve *ED* indicates the value of the benefit (the benefit for users as evaluated by the market) for a marginal user under the condition that the expected penetration rate is equal to the actual penetration rate. However, if all users expect a penetration rate higher than the actual penetration rate, demand curve *ED* will shift upward.[6] It has been discussed that 'history' and 'expectation' play an important role in the selection of an equilibrium point in an economy where positive feedback exists (Krugman, 1991; Matsuyama, 1991). The good equilibrium will actually be realized when economic entities expect 'good equilibrium' to be the final equilibrium among several other possible equilibriums, and these entities then act according to such an expectation. The government will thus be able to politically realize a desirable equilibrium point by controlling the expectations of the private sector. Such public announcements thus grant the government the ability to decrease not only the level of the critical mass but also that of the Pigouvian subsidy, thereby reducing the financial cost of realizing the optimal penetration rate.

5.5 Conclusions

In this chapter we examined certain features of the benefits of ITS technologies, taking VICS and safety-related ITS technologies as examples. We then explained the notion of the *optimal penetration rate* and discussed policies for achieving it.

The conclusions of our analyses are as follows. There are systems with certain network externality features that have no analogue among other goods and services in the IT industry: (a) systems with negative network externality, and (b) systems for which the sign of the network externality changes from positive to negative with increasing penetration rate. Second, in order to realize the optimal penetration rate for a system with positive network externality, such as the car-to-car system, it is necessary to distribute free systems at a level slightly exceeding the critical mass, as well as to provide a Pigouvian subsidy. In this case, the government can decrease both the level of the critical mass and the Pigouvian subsidy by modifying drivers' expectations for the penetration rate, thus reducing the financial cost of realizing the optimal penetration rate.

Notes

1. In describing the features of each ITS in this section, we referred to the websites of the Japanese Ministry of Land, Infrastructure and Transport (http://

www.mlit.go.jp/road/ITS/), the Japanese National Police Agency (http://www.npa.go.jp/), and Honda Motor Co., Ltd (http://world.honda.com/). We accessed these sites on November 30, 2010.

2. There will also be a decision-making benefit for users, which allows drivers to make appropriate decisions about whether or not to travel (Emmerink *et al.*, 1996; Zang and Verhoef, 2006).
3. Levinson (2003) summarized the results of previous studies concerning the benefits for equipped drivers and the improvement in system performance.
4. Throughout this chapter we have chosen the sense of left and right turns to place our discussions in a context most familiar to US and European readers; thus, the type of collision considered here, between a left-turning vehicle and a vehicle passing through an intersection, is in fact characteristic of *right-turning* vehicles in Japan (where automobiles travel on the left side of the road, not the right side as in the US and continental Europe).
5. Although we will not discuss this point in detail here, the car-to-car systems, unlike the car-to-infrastructure systems, exhibit no difference in accident reduction rate (at least in theory) between the dual-info and single-info cases.
6. In this case, the term 'expected equilibrium demand curve' is no longer technically accurate.

References

Arnott, R., Palma, A., and Lindsey, R. (1991) 'Does Providing to Drivers Reduce Traffic Congestion?', *Transportation Research Part A*, 25(5): 309–18.

Ben-Akiva, M., Palma, A., and Kaysi, I. (1991) 'Dynamic Network Models and Drivers Information Systems', *Transportation Research Part A*, 25(5): 251–66.

Emmerink, R. H. M., Nijkamp, P., and Rietvieid, P. (1994) 'The Economics of Motorist Information System Revisited', *Transport Review*, 14(4): 363–88.

Emmerink, R. H. M., Axhausen, K. W., Nijkamp, P., and Rietveld, P. (1995) 'Effects of Information in Road Transport Networks with Recurrent Congestion', *Transportation*, 22(1): 21–53.

Emmerink, R. H. M., Verhoef, E. T., Nijkamp, P., and Rietveld, P. (1996) 'Endogenising for Information in Road Transport', *Annals of Regional Science*, 30(2): 201–22.

Huang, H. J. and Li, Z. C. (2007) 'A Multiclass, Multicriteria Logit-based Traffic Equilibrium Assignment under ATIS', *European Journal of Operational Research*, 176(3): 1464–77.

Katz, M. L. and Shapiro, C. (1985) 'Network Externalities, Competition, and Compatibility', *American Economic Review*, 75(3): 424–40.

Katz, M. L. and Shapiro, C. (1994) 'Systems Competition and Effects', *Journal of Economic Perspectives*, 8(2): 93–115.

Koutsopoulos, H. N. and Lotan, T. (1990) 'Motorist Information Systems and Recurrent Traffic Congestion: Sensitivity of Expected Result', *Transportation Research Record*, 1281: 145–58.

Krugman, P. R. (1991) 'History versus Expectations', *Quarterly Journal of Economics*, 106(2): 651–67.

Kulmala, R. (2010) 'Ex-ante Assessment of the Safety Effects of Intelligent Transport Systems', *Accident Analysis & Prevention*, 42(4): 1359–69.

Levinson, D. (2003) 'The Value of Advanced Information Systems for Route Choice', *Transportation Research Part C*, 11(1): 75–87.

Libenstein, H. (1950) 'Bandwagon, Snob and Veblen Effects in the Theory of Consumer's Demand', *Quarterly Journal of Economics*, 64(2): 183–207.

Lo, H., Hickman, M., Ran, B., Larson, J., and Weissenberger, S. (1995) 'Route Guidance and Planning: Potential Benefits and for Public-private Partnerships', *Proceedings of the ITS America Fifth Annual Conference*, 2: 767–75.

Mahmassani, H. and Jayakrishnan, R. (1991) 'System Performance and Response under Real-time Information in a Congested Traffic Corridor', *Transportation Research Part A*, 25(5): 293–307.

Matsuyama, K. (1991) 'Increasing Returns, Industrialization, and Indeterminacy of Equilibrium', *Quarterly Journal of Economics*, 106(2): 617–50.

Rohlfs, J. H. (1974) 'A Theory of Independent Demand for a Communications Service', *Bell Journal of Economics and Management Science*, 5(1): 16–37.

Spyropoulou, I., Penttinen, M., Karlaftis, M., Vaa, T., and Golias, J. (2008) 'ITS Solutions and Accident Risks: Prospective and Limitations', *Transport Reviews*, 28(5): 549–72.

Vaa, T., Penttinen, M., and Spyropoulou, I. (2007) 'Intelligent Transport and Effects on Road Traffic Accidents: State of the Art', *IET Intelligent Transport Systems*, 1(2): 81–8.

Wardrop, J. (1952) 'Some Theoretical Aspects of Road Research', *Proceedings of the Institute of Civil Engineers*, 1(Part 2): 325–78.

Yang, H. (1999) 'Evaluating the Benefit of a Combined Route and Road Pricing System in a Network with Recurrent Congestion', *Transportation*, 26(3): 299–322.

Zang, R., and Verhoef, E. T. (2006) 'A Monopolistic Market for Traveler Information Systems and Road Efficiency', *Transportation Research Part A*, 40(5): 424–43.

6
Optimal Market Penetration Rates of VICS

Hiroaki Miyoshi and Masayoshi Tanishita

6.1 Introduction

VICS (Vehicle Information and Communication System) is a digital data communication system which provides prompt and useful information on current traffic conditions – such as traffic congestion, accidents, or link-travel time – to drivers via car navigation equipment. VICS was described in detail in Chapter 5.

In this chapter, we analyze the optimal penetration rate of VICS units, and we use simulation models to estimate the taxes and/or subsidies required to attain such a level.

Several models of market penetration rates for traveler information systems have been established (see, for example, Emmerink *et al.*, 1996, 1998; Huang *et al.*, 2008; Li 2004; Li *et al.*, 2003; Yang, 1998, 1999; Yang and Meng, 2001; Yin and Yang, 2003; Zang and Verhoef, 2006). Among these references, Emmerink *et al.* (1996) and Zang and Verhoef (2006) discussed tax and subsidy policies for traveler information systems from the perspective of economic welfare, as we do in this chapter.

One of the most important differences between the aforementioned works and this chapter is the object of our penetration rate analysis. The penetration rates analyzed in the above studies were the rates of utilization of traffic information on specific origin-destination pairs and hypothetically determined road networks. In contrast, our focus will be on the penetration rate of VICS-enabled onboard car navigation devices that receive traffic information from VICS (referred to hereafter as *VICS units*). This is because, if a driver purchases a VICS unit, she will be able to receive traffic information for free, and it makes no sense to consider the problem of penetration rates for individual routes. In analyzing the penetration level of VICS units, two points of view that differ from previous studies will be important. The first point is the network externality of VICS units and the second is the problem of dynamic stability for penetration levels attributed to its network externality.

In this chapter, we first develop a simulation model in accordance with the theoretical concepts discussed in Chapter 5. Then, in Section 6.3, we calculate the market penetration rate, the optimal penetration rate, the dynamic stability of the optimal penetration rate, and the amount of taxes and/or subsidies required to realize this optimal penetration rate.

6.2 Simulation model

In this section we present our simulation model.

6.2.1 Primary assumptions and framework of the simulation model

For the purposes of these simulations, we consider a fictitious locale in which travel distances, travel speeds, and other travel-related quantities are set equal to their nationwide average values in Japan. The primary assumptions and modeling framework underlying our simulations are as follows.

a) Object of the simulation.

Among the benefits of VICS discussed in Chapter 5, the object of this simulation will be to quantify the travel time savings benefit enjoyed by users, as well as that enjoyed by non-users.

b) Vehicle classes.

We have categorized vehicles into four classes: passenger cars, mini-sized trucks, small-sized trucks, and ordinary-sized trucks. We have set different travel distance distributions and travel times for each vehicle class in order to calculate the travel time savings benefit. In general we have referred to Japanese statistical data for 2007 and 2008 in setting travel distance distributions and travel times.

c) Relationship between the VICS unit penetration rate and the travel time reduction rate.

We assume that reductions in travel time will be realized only when users drive on open roads in urban areas,[1] and will not be realized when users drive in other areas.

Regarding the relationship between the VICS unit penetration rate and the travel time reduction rate, we note that almost all VICS units currently in use are so-called 'one-media receiver' units, which receive only FM multiplex broadcast signals, and we thus assume that the relationship between VICS unit penetration rate and travel time reduction rate is as illustrated in Figure 5.1 of Chapter 5. In our simulations we have used several values for α (the travel time reduction rate for users at a penetration rate of 0 per cent) and β (the travel time reduction rate for users at a penetration rate of 100 per cent) as well as for the curvatures of the curves plotted in Figure 5.1.

It should be noted that Figure 5.1 plots the *expectation* value of the travel time reduction rate, while the actual values of the travel time reduction rate will be distributed stochastically.

d) Decision-making involved in purchasing a VICS unit.

We assume that each individual has complete information on the relationship between the VICS unit penetration rate and the expected value of travel time savings. Each individual will purchase a VICS unit when the expected benefit (see the part labeled 'Benefit for users as evaluated by the market' in Figure 5.4 in Chapter 5) at the expected penetration rate exceeds the cost of purchasing a VICS unit.

e) Change in traffic volume due to penetration of VICS units.

As VICS reduces the generalized cost of travel, it is reasonable to assume that the penetration of VICS units will increase traffic volume. In this chapter, however, in order to simplify the model, we conduct our simulations under the assumption that VICS does not affect the volume of traffic.

6.2.2 Model

The simulation model and parameter values we use are as follows.

a) Distribution of travel time costs in urban areas.

In Chapter 5, we examined the benefits of using VICS under the assumption that time spent traveling and the value of time are equal for all drivers. In this simulation, in contrast, we assume that the monthly cost of time spent traveling (the product of the time spent traveling and the value of time) in urban areas is log-normally distributed. We denote by x_j the monthly cost of time spent traveling in vehicles of type j, where $j \in \{c, m, s, o\}$; here c denotes passenger cars, m mini-sized trucks, s small-sized trucks, and o ordinary-sized trucks. We assume different distributions for different types of vehicle, as we now describe.

Passenger cars

We assume that the distance traveled by a passenger car ($j = c$) and the value of time for individuals driving passenger cars are independent and log-normally distributed. Under these assumptions, the travel time cost for a passenger car in urban areas is log-normally distributed as follows:

$$d_c(x_c) = \frac{1}{\sigma_{x_c} x_c \sqrt{2\pi}} \exp\left[-\frac{(\ln x_c - \mu_{x_c})^2}{2\sigma_{x_c}^{\ 2}}\right], \tag{6.1}$$

where $\mu_{x_c} = \eta_{d,c} + \ln \lambda_c - \ln \nu + \eta_{w,c}$ and $\sigma_{x_c} = \sigma_{d,c_i} + \eta_{w,c}$, and where

$\eta_{d,c}$ and σ_{d,c_i} are calculated from the national average and median monthly travel distance,[2] under the assumption that the monthly travel distance is log-normally distributed,

λ_c is the national average of the fraction of total travel conducted on open urban roads,[3] which we assume to be identical for all passenger cars,

ν is the average speed of travel on open roads in urban areas,[4] and

$\eta_{w,c}$ and $\sigma_{w,c}$ are calculated from the national average and median value of time under the assumption that the value of time is log-normally distributed. We have calculated the average and median of the value of time based on the distribution of annual income[5] and the average number of hours worked annually[6] nationwide.

Table 6.1 shows values for each parameter calculated in accordance with the methods discussed above.

Mini-sized trucks, small-sized trucks and ordinary-sized trucks

We assume that the travel distance for each type of truck ($j \in \{m, s, o\}$) is log-normally distributed. Under this assumption, the travel time cost in urban areas is also log-normally distributed as follows:

$$d_j(x_j) = \frac{1}{\sigma_{x_j} x_j \sqrt{2\pi}} \exp\left[-\frac{(\ln x_j - \mu_{x_j})^2}{2\sigma_{x_j}^{\ 2}}\right], \quad j \in \{m, s, o\}, \tag{6.2}$$

where $\mu_{x_j} = \eta_j + \ln \lambda_j - \ln \nu + \ln w_j$, and where

η_j and σ_{x_j} are calculated from the national average and median values of monthly travel distance,[7] under the assumption that the monthly travel distance for vehicle j is log-normally distributed,

λ_j is the national average of the fraction of total travel distance conducted on open urban roads,[8] which is assumed to be the same for all vehicles of a given type,

Table 6.1 Values used to determine parameters in the distribution of travel time costs for passenger cars

μ_{x_c}	$\eta_{d,c}$	$\eta_{w,c}$	λ_c	v	σ_{x_c}	σ_{d,c_i}	$\sigma_{w,c}$
9.518	5.665	7.8542	0.415	22.676	1.476	0.894	0.583

v is the speed of travel (which is the same as that for passenger cars), and

w_j is the value of time. We have employed the values listed in the *Cost–Benefit Analysis Manual* published by the Japanese Ministry of Land, Infrastructure and Transport,[9] under the assumption that the value of time is the same for all vehicles of a given type. Although we have assumed that the value of time is log-normally distributed for passenger cars, we take the value of time to be a constant for each type of truck.

Table 6.2 shows values for each parameter calculated in accordance with the methods discussed above.

b) Travel time reduction rate.

We obtain the following equation for the expected value $rt_{with}(p)$ of the travel time reduction rate for users:

$$rt_{with}(p) = 2(\alpha - \beta)(1 - 2\theta_{with})p^2 + (\beta - \alpha)(1 - 4\theta_{with})p - \alpha, \alpha > \beta. \quad (6.3)$$

In this formulation, the travel time will be reduced by 100α per cent at a penetration rate of 0 per cent, by $100(\alpha - (\alpha - \beta)\theta_{with})$ per cent at a penetration rate of 50 per cent, and by 100β per cent at a penetration rate of 100 per cent.

On the other hand, we obtain the following quadratic expression for the expected value $rt_{with}(p)$ of the travel time reduction rate for non-users:

$$rt_{without}(p) = 2\beta(2\theta_{without} - 1)p^2 + \beta(1 - 4\theta_{without})p. \quad (6.4)$$

In this formulation, the travel time will not be reduced at a penetration rate of zero, but will be reduced by $100\beta\theta_{without}$ per cent at a penetration rate of 50 per cent and by 100β per cent at a penetration rate of 100 per cent (the same reduction rate as for users).

θ_{with} and $\theta_{without}$ are parameters which determine respectively the curvatures of $rt_{with}(p)$ and $rt_{without}(p)$. Here we use a base case in which

Table 6.2 Values used to calculate parameters in the distribution of travel time costs for trucks

	μ_{x_j}	η_j	λ_j	ν	w_j	σ_{x_j}
Mini-sized trucks ($j = m$)	9.842	5.981	0.375	22.676	2,875	1.233
Small-sized trucks ($j = s$)	10.574	6.713	0.375	22.676	2,875	1.023
Ordinary-sized trucks ($j = o$)	12.380	8.256	0.364	22.676	3,851	0.464

Table 6.3 Values of curvature parameters for the base case and the comparison cases

	Value of θ_{with} for $rt_{with}(p)$		
Value of $\theta_{without}$ for $rt_{without}(p)$	**1/4 (concave)**	**2/4 (linear)**	**3/4 (convex)**
1/4 (convex)	Case 8	Case 4	Case 6
2/4 (linear)	Case 2	Base case	Case 1
3/4 (concave)	Case 7	Case 3	Case 5

Note: 'Concave' indicates that the function in question is concave to the original point. 'Convex' means that the function is convex to the original point.

$rt_{with}(p)$ and $rt_{without}(p)$ are linear, as well as eight other cases for comparison with the base case. Table 6.3 shows values of θ_{with} and $\theta_{without}$ for the base case and for the comparison cases. In the interval $0 \leq p \leq 1$, $rt_{with}(p)$ is a monotonically increasing function and $rt_{without}(p)$ is a monotonically decreasing function for all cases.

c) Benefit enjoyed by users.

The expected value of the benefit derived by a user of each vehicle type at a penetration rate of p can be expressed as the product of two factors: (a) the margin between the travel time reduction rate for users and that for non-users, and (b) the travel time cost:

$$ben_{j,with}(p,x_j) = x_j(rt_{without}(p) - rt_{with}(p)), \quad j \in \{c,m,s,o\}. \tag{6.5}$$

Curve of expected equilibrium demand

At a given penetration rate p, the expected value of the benefit for marginal users is the same for vehicles of all types. As is clear from Eq. (6.5), this means that, at a given penetration rate p, the travel time cost for marginal users is the same among vehicles of all types.

The curve of expected equilibrium demand indicates the value of the benefit as evaluated by the market for marginal users under the condition that the expected penetration rate is equal to the actual penetration rate. If the travel time cost is $x^* = x_c^* = x_m^* = x_n^* = x_o^*$, then the expected equilibrium demand curve ED satisfies the following equation:

$$F(p, x^*) = \frac{1}{\bar{n}} \sum_j \bar{n}_j \int_{x^*}^{\infty} d_j(x_j)\, dx_j - p = 0, \quad j \in \{c, m, s, o\}, \tag{6.6}$$

where $\bar{n} = \sum_j \bar{n}_j$.

If we assume that Eq. (6.6) implicitly defines the function $x^* = h(p)$, then ED at a penetration rate of p can be expressed as the function of p as follows:

$$\begin{aligned} ED(p) &= ben_{c,with}(p, x^*)|_{x^*=h(p)} \\ &= ben_{m,with}(p, x^*)|_{x^*=h(p)} \\ &= ben_{s,with}(p, x^*)|_{x^*=h(p)} \\ &= ben_{o,with}(p, x^*)|_{x^*=h(p)}. \end{aligned} \tag{6.7}$$

However, Eq. (6.6) cannot be explicitly solved for x^*. We thus solve for x^* numerically, using Newton's method, at fixed values of p in Eq. (6.6). We also derive a value for ED at a penetration rate p by plugging p and the corresponding value of x^* into Eq. (6.7).

Table 6.4 lists, for each type of vehicle, the number $\bar{n}_j$ of vehicles used in the simulation.[10]

Total benefit for all users

We can assume that the probability distribution of the quantity $(rt_{without}(p) - rt_{with}(p))$, the difference between the travel time reduction rate for users and that for non-users, is independent of the probability distribution of the travel time cost for each type of vehicle. The *total* benefit enjoyed by all vehicles of type j, $sben_{j,with}$, is equal to the definite integral, from the travel time cost for marginal users (as obtained by solving Eq. (6.6) numerically) to ∞, of the product of three factors: (a) the expected value (6.5) of the benefit for vehicle type j, (b) the travel time cost probability density function for vehicles of type j (as given

Table 6.4 Number of vehicles for each type at the end of March 2008

Type of vehicle	Number $\bar{n}_j$
Passenger car $j = c$	57,551,248
Mini-sized truck $j = m$	9,380,627
Small-sized truck $j = s$	4,283,313
Ordinary-sized truck $j = o$	2,445,264

Source: Japanese Automobile Inspection & Registration Association.

by Eqs. (6.1) and (6.2)), and (c) the number of vehicles $\bar{n}_j$ of type j. The overall aggregate benefit enjoyed by all users of all vehicle types is the sum of the total benefit for each of the four vehicle types:

$$sben_{with}(p,x^*) = \sum_j sben_{j,with}(p, x^*)$$
$$= \sum_j \bar{n}_j \int_{x^*}^{\infty} ben_{j,with}(p,x_j)\, d_j(x_j)dx_j. \quad (6.8)$$

$sben_{j,\, with}$ is equivalent to the quantity labeled 'Benefit for users as evaluated by the market' in Figure 5.4.

Similarly, the marginal value $msben_{with}$, the additional benefit enjoyed by all vehicles when the number of vehicles increases by one, may be expressed in the form:

$$msben_{with}(p, x^*) = \frac{1}{\bar{n}} \sum_j \left(\frac{\partial}{\partial p} sben_{j,with}(p, x^*) - \frac{F_p}{F_{x^*}} \frac{\partial}{\partial x^*} sben_{j,with}(p, x^*) \right), \quad (6.9)$$

where $F_p = \dfrac{\partial}{\partial p} F(p, x^*)$, $F_{x^*} = \dfrac{\partial}{\partial x^*} F(p, x^*)$.

To obtain values for Eqs. (6.8) and (6.9), we use Newton's method to solve Eq. (6.6) for x^* at a fixed value of p, then substitute these values of $x^*$and p into Eqs. (6.8) and (6.9).

d) Benefit enjoyed by all vehicles.

At a penetration rate of p, the expected value of the benefit that is *not* reflected in the market for marginal users and the expected value of the benefit for marginal non-users is given by the expression $-x_j\, rt_{without}(p)$.

Then the quantity $sben_{j,together}$, the total benefit enjoyed by all vehicles of type *j*, is equal to the definite integral, from 0 to ∞, of the product of three factors: (a) the expected value of the benefit that is *not* reflected in the market for marginal vehicle of type *j*, (b) the travel time cost probability density function for vehicles of type *j* (given by Eqs. (6.1) and (6.2)), and (c) the number $\bar{n}_j$ of vehicles of type *j*. The overall aggregate benefit $sben_{together}$ for all vehicles of all types is the sum of the total benefit for each of the four vehicle types:

$$\begin{aligned} sben_{together}(p) &= \sum_j sben_{j,together}(p) \\ &= -\sum_j \bar{n}_j \, rt_{without}(p) \int_0^{\infty} x_j \, d_j(x_j) \, dx_j. \end{aligned} \tag{6.10}$$

Similarly, the marginal value $msben_{together}$, the additional benefit enjoyed by all vehicles when the number of vehicles increases by one, is

$$msben_{together}(p) = \frac{1}{\bar{n}} \frac{d}{dp} sben_{together}(p). \tag{6.11}$$

e) Social benefit and marginal social benefit.

The social benefit *sumben* is the sum of the total benefit enjoyed by all users (Eq. (6.8)) and the total benefit enjoyed by all vehicles (Eq. (6.10)):

$$sumben(p, x^*) = sben_{with}(p, x^*) + sben_{together}(p). \tag{6.12}$$

On the other hand, from Eqs. (6.9) and (6.11), the marginal social benefit *msumben* may be expressed in the form:

$$msumben(p, x^*) = msben_{with}(p, x^*) + msben_{together}(p). \tag{6.13}$$

f) Social cost and marginal social cost.

The social cost of VICS is the sum of the cost of infrastructure improvements, such as vehicle detectors installed on roads, and the cost of supplying VICS units. We assume that the former is a fixed cost (an initial cost), while the latter is a variable cost that varies with the number of VICS units. In addition, we assume that the production function of VICS units displays constant returns to scale. Under these assumptions, the marginal social cost of a VICS unit is equal to the price of a VICS unit.

In this simulation, as discussed above, we calculate the benefit of using VICS for a one-month period. We must thus take the price (the marginal social cost) to be the usage fee for one month. If we assume that the price of a VICS unit is 100,000 yen, that its lifetime is 10 years, and that its discount rate is 7 per cent, we arrive at a monthly usage fee of 1,109 yen. In this chapter we take this value as the marginal social cost.

6.2.3 Optimal penetration rate of VICS units and its dynamical stability

a) Optimal penetration rate of VICS units.

The optimal penetration rate must satisfy the following first-order (Eq. (6.14)) and second-order (Eq. (6.15)) conditions:

$$msumben\,(p\,,x^{*}) = c\,, \tag{6.14}$$

$$\frac{d}{dp}\,msumben\,(p\,,x^{*}) \leq 0. \tag{6.15}$$

b) Dynamic stability of the optimal penetration rate.

The optimal penetration rate derived from Eqs. (6.14) and (6.15) is not the same as the market penetration rate, which is the point of intersection of the expected equilibrium demand curve, *ED*, and the marginal social cost. Thus, as discussed in Chapter 5, we need Pigou taxes or subsidies to realize the optimal penetration rate. The optimal penetration rate will then be realized as the point of intersection of *ED* and the supply price including such taxes or subsidies.

As discussed above, *ED* is plotted under the condition that the expected penetration rate equals the actual market penetration rate. When there is a discrepancy between the actual penetration rate and the expected penetration rate, it is a non-trivial matter to determine whether the market penetration rate converges to a point on the expectation equilibrium demand curve.

Here we assume that all individuals have the same expectation for the penetration rate of VICS units, p^e. If we denote the supply price, including taxes or subsidies (the market price of VICS units), by c', then the travel time cost x' (a common value for all types of vehicle) for marginal VICS unit users under this expected penetration rate satisfies the following equation:

$$G\,(p^e, x') = ben_{c,with}\,(p^e, x') - c' = 0. \tag{6.16}$$

If we assume that Eq. (6.16) implicitly defines the function $x'=l(p^e)$, then the actual penetration rate p that will be realized can be expressed as a function of the expected penetration rate p^e:

$$p = k\,(x')|_{x'=l(p^e)} = \frac{1}{\bar{n}} \sum_j \bar{n}_j \int_{x'}^{\infty} d_j\,(x_j)\,dx_j\,|_{x'=l(p^e)}. \tag{6.17}$$

Then, if we assume that all players take the penetration rate for the previous term as the expected penetration rate of the present term ($p^e = p_{-1}$), the slope of the phase diagram may be expressed as a function of the penetration rate in the previous term:

$$\frac{dp}{dp_{-1}} = -\frac{G_{p^e}}{G_{x'}} \frac{d}{dx'} k\,(x')|_{x'=l(p^e)}, \quad p^e = p_{-1}. \tag{6.18}$$

If the absolute value of Eq. (6.18) is lower than 1 at the optimal penetration rate, then the penetration rate will converge to a point on *ED*, and the optimal penetration rate will be dynamically stable, assuming that the expected penetration rate is in the neighborhood of the optimal penetration rate. Otherwise, the penetration rate will not converge to a point on *ED* and will be a dynamically unstable penetration rate.

We have derived values of Eqs. (6.17) and (6.18) by fixing a value for p^e in Eq. (6.16) and by inserting the corresponding numerical solution for x' into Eqs. (6.17) and (6.18).

6.3 Simulation results

Using the methods discussed above, we have conducted simulations to assess the benefits of VICS, the market penetration rate, the optimal penetration rate, and the rates of taxes and/or subsidies needed to realize the optimal penetration rate. In this section, unless otherwise stated, we will present results for the base case (the case in which both $rt_{with}(p)$ and $rt_{without}(p)$ are linear).

6.3.1 Expected equilibrium demand curve, marginal value of benefit for all users, and marginal social benefit

First, to confirm the shapes of, and the relative positioning among, the expected equilibrium demand curve (*ED*), the curve of marginal value of benefit for all users (*MUB*), and the curve of marginal social benefit (*MSB*), we have drawn these curves using the values (0.1, 0.05) for the travel time reduction rate (α, β). The result is shown in Figure 6.1.

In terms of externality, this figure differs in two ways from the corresponding figure drawn for ITS technologies with positive network externality (Figure 5.10 in Chapter 5). One is that the expected equilibrium demand curve in Figure 6.1 has negative slope. The other is that the curve of marginal value of benefit for all users, *MUB*, lies below the expected equilibrium demand curve *ED*. These two features arise from the negative network externality of VICS.

The intersection of the price curve *C* (the monthly usage fee) and *MSB* is the optimal penetration rate. In this case, the optimal penetration rate is realized at a penetration rate of 28.9 per cent.[11] However, at this penetration rate, *ED* lies above *MSB*, indicating that the penetration rate exceeds the optimal market penetration rate. In order to realize the penetration rate of 28.9 per cent in the market, the cost to users must be

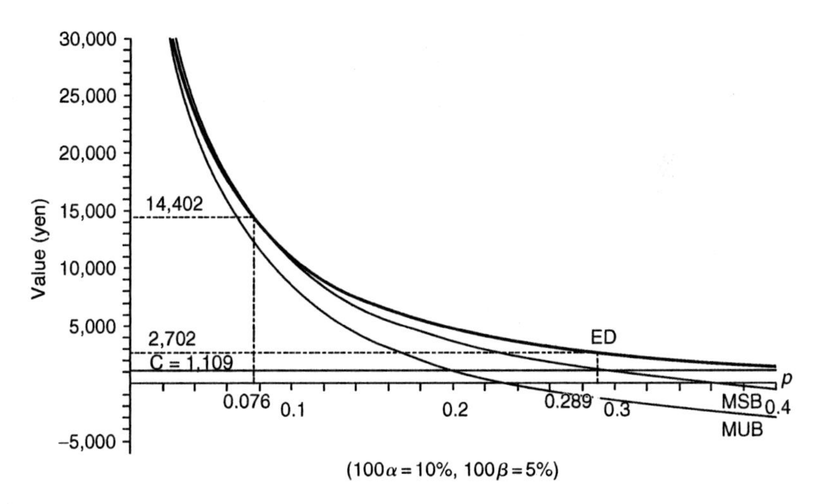

Figure 6.1 Expected equilibrium demand curve (*ED*), marginal value of benefit for all users (*MUB*), and marginal social benefit (*MSB*)

2,702 yen. To realize the optimal penetration rate we must thus impose a tax of 1,593 yen (the difference between 2,702 yen and the marginal social cost of 1,109 yen) on every VICS unit.

6.3.2 Optimal penetration rate, market penetration rate, and level of taxes and/or subsidies needed to achieve the optimal penetration rate at various travel time reduction rates

Next, we have calculated the optimal penetration rate and the market penetration rate for various values of the travel time reduction rate parameters (α, β), with results as shown in Figure 6.2. The four thin lines in Figure 6.2 show how the market penetration rate changes with the value of β, for four different values of α (0.05, 0.10, 0.15, and 0.20).[12] The four bold lines show how the optimal penetration rate changes with the value of β at the same four values of α.

We first note that the market penetration rate is independent of β, because, in the base case, the margin between the travel time reduction rate for users and that for non-users does not depend on β. Comparison cases 5 and 8 are cases in which the market penetration rate does not depend on β. In contrast, in comparison cases 2, 3, and 7, the market

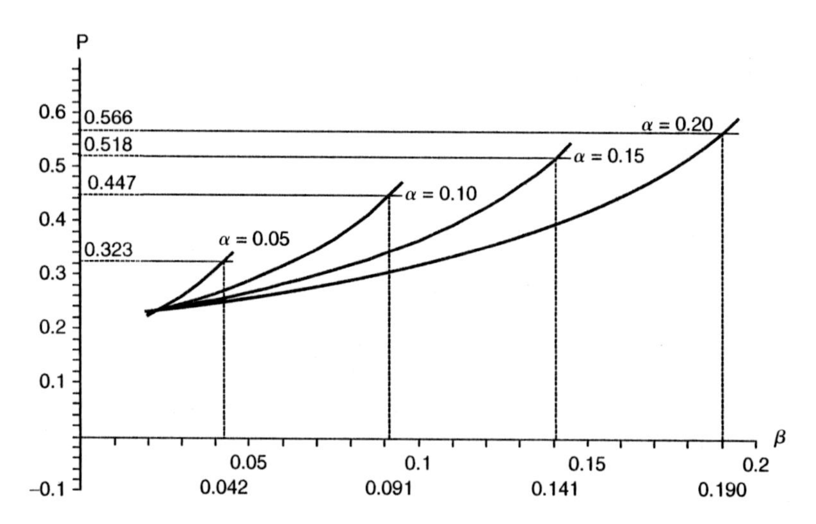

Figure 6.2 Optimal penetration rate and market penetration rate at various values of the travel time reduction rate parameters

Note: Thin curves show market penetration rates, while thick curves show optimal penetration rates, at various values of the travel time savings parameters (α, β).

penetration rate decreases with increasing β, while in comparison cases 1, 4, and 6 the market penetration rate increases with increasing β. On the other hand, the market penetration rate always increases with increasing α at fixed β. This is because *ED* shifts to the right as the value of α increases.

Next, the optimal penetration rate increases with increasing β at constant α. This is because the benefit enjoyed by both users and non-users (the sum of the quantities 'Benefit for users that is not reflected in the market' and 'Benefit for non-users' in Figure 5.4) increases with increasing β. In contrast to the market penetration rate, the optimal penetration rate *decreases* with increasing α at constant β. This is because the effect of negative network externality increases as the gap between α and β becomes larger.

Considering next the relationship between the market penetration rate and the optimal penetration rate, we note that these quantities are equal at the points $(\alpha, \beta, p) = (0.05, 0.042, 0.323)$, $(0.10, 0.091, 0.447)$, $(0.15, 0.141, 0.518)$, and $(0.20, 0.190, 0.566)$. If the value of β exceeds the β values of these points at the corresponding α value, the optimal penetration rate will exceed the market penetration rate and we will have to subsidize VICS units to realize the optimal penetration rate. On the other hand, if the value of β is below the β values of these points at the corresponding α value, the optimal penetration rate will fall below the market penetration rate. In this case we will have to impose taxes on VICS units to realize the optimal penetration rate.

Figure 6.3 shows the level of taxes or subsidies required to realize the optimal penetration rate as a function of β, at the four values of α considered above. Positive values of the vertical coordinate correspond to taxes, while negative values correspond to subsidies. This figure clearly illustrates that, unless the value of β is similar to the value of α, we will have to impose taxes on VICS units to realize the optimal penetration rate.

We have calculated the value of β at which the market penetration rate and the optimal penetration rate become equal for 37 values of α, ranging from a minimum value of 0.02 to a maximum value of 0.20 in steps of 0.005. The heavy line in Figure 6.4 was drawn by connecting these points with a straight line. On the other hand, the thin line in Figure 6.4 is the line $\alpha = \beta$. In the region below the heavy line, taxes must be imposed to realize the optimal penetration rate, while, at points lying between the heavy and thin lines, subsidies must be provided to realize the optimal penetration rate. Like Figure 6.3, this figure again

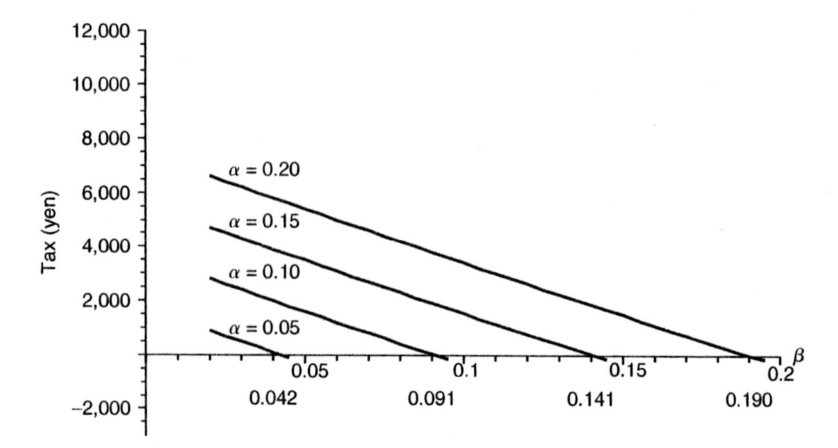

Figure 6.3 Taxes or subsidies required to realize the optimal penetration rate as a function of the travel time reduction rate parameters (1)

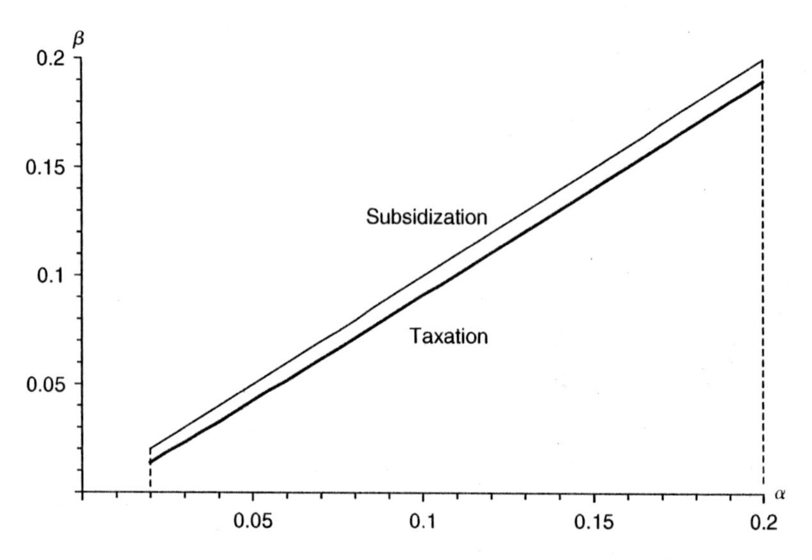

Figure 6.4 Taxes or subsidies required to realize the optimal penetration rate as a function of the travel time reduction rate parameters (2)

illustrates that, unless the values of β and α are similar, we will need to impose taxes to realize the optimal penetration rate.

On the other hand, another way to think about the content of these figures is to note that points (α, β) lying on the bold line are points at which the extent to which the impact of the negative network externality reduces the marginal benefit for all users (the difference between *MUB* and *ED*) just equals the marginal value of the benefit enjoyed by users and non-users alike (the difference between *MSB* and *MUB*).

In this simulation, we have not measured the benefit of car navigation systems belonging to individual users. However, this benefit does not exhibit negative and positive network externality. If we assume that the users who, we predict in this chapter, will derive the greatest benefits from VICS are also the users who place the highest value on car navigation systems, then there should be no change in the difference between the curve of marginal benefit for all users (*MUB*) and the curve of marginal social benefit (*MSB*) even *if* we include the effects of car navigation systems. Thus, accounting for the impact of car navigation systems should cause all three curves in Figure 6.1 to shift upward without changing their relative positioning. This enhances the need for the imposition of taxes to realize the optimal penetration rate.

The results discussed above are from simulations of the base case. Table 6.5 shows, for each of the comparison cases, the values of (α, β) at which the market penetration rate equals the optimal penetration rate. For example, in comparison case 1, the market penetration rate coincides with the optimal penetration rate at the point (α, β) = (0.05, 0.044), and the market penetration rate at this rate is 31.6 per cent. In all cases the market

Table 6.5 Values of the travel time reduction rate parameters at which the market penetration rate coincides with the optimal penetration rate in the various comparison cases

	α=0.05		α=0.1		α=0.15		α=0.2	
Standard case	0.042	0.323	0.091	0.447	0.141	0.518	0.190	0.566
Comparative case 1	0.044	0.316	0.092	0.441	0.140	0.512	0.189	0.561
Comparative case 2	0.040	0.335	0.091	0.454	0.141	0.523	0.192	0.570
Comparative case 3	0.038	0.282	0.085	0.377	0.134	0.432	0.183	0.470
Comparative case 4	0.045	0.375	0.094	0.516	0.143	0.592	0.193	0.389
Comparative case 5	0.040	0.270	0.087	0.365	0.135	0.422	0.184	0.461
Comparative case 6	0.046	0.372	0.093	0.511	0.141	0.587	0.186	0.643
Comparative case 7	0.034	0.303	0.084	0.392	0.134	0.444	0.184	0.479
Comparative case 8	0.045	0.426	0.095	0.564	0.145	0.636	0.194	0.455

Note: Each entry in the table lists, for the given value of α, the pair (β, p) such that (α, β) are the values of the travel time reduction rate parameters at which the market penetration rate and optimal penetration rates are equal to each other and to p. For example, in comparison case 1, the market penetration rate coincides with the optimal penetration rate when $\alpha = 0.05$ and $\beta = 0.044$, and the market penetration rate in this case is 31.6 per cent.

penetration rate coincides with the optimal penetration rate when α and β are similar in value. We can thus conclude that, in the comparison cases just as in the base case, the imposition of taxes will be necessary to realize the optimal penetration rate unless the value of β is similar to the value of α.

6.3.3 Dynamical stability

Next let us consider the dynamical stability of the optimal penetration rates discussed above. The four curves in Figure 6.5 show the slopes of the phase lines at the optimal penetration rate, versus the value of α, for the four values of β considered previously. These curves have been obtained by substituting the optimal penetration rate for p^e in Eq. (6.18).

This figure indicates whether or not the imposition of taxes or subsidies causes the market penetration rate to converge to the optimal penetration rate, assuming the expected penetration rate is in the neighborhood of the optimal penetration rate.

In Figure 6.5, for all cases, the slopes of the phase lines are negative and lower than 1 in absolute value. This indicates that, when the expected penetration rate is in the neighborhood of the optimal penetration rate, the market penetration rate will converge to the optimal penetration rate.

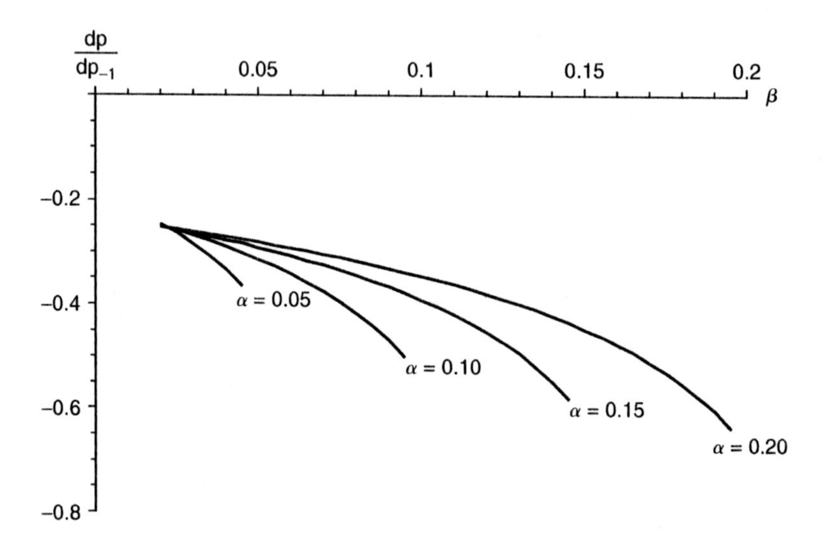

Figure 6.5 Dynamical stability of the optimal penetration rate as a function of the travel time reduction rate parameters

6.4 Conclusions

In this chapter we have used a simulation model to investigate the optimal penetration rate of VICS units and the quantity of taxes or subsidies required to realize that rate. Although the analysis depends strongly on our assumptions, the results suggest that it will be necessary to impose taxes on VICS units to realize the optimal penetration rate. In the future we intend to improve and extend this research in the following two directions.

The first challenge is to conduct simulations of market penetration rate and optimal penetration rate, as well as taxes or subsidies required to realize the optimal rate, under the assumption that the sign of the network externality changes from positive to negative with increasing penetration rate, as in the case depicted in Figures 5.2 and 5.3 in Chapter 5. Such simulations will be helpful in investigating tax and subsidy policies for the case in which the three-media receiver VICS unit or the Probe-Car Information System, which collects traffic information by using vehicles as sensors interconnected through a radio network, becomes mainstream.

The second challenge is to refine our assessments of VICS benefits and optimal penetration rate. In this simulation, we have chosen the travel time savings benefit as the object of measurement. However, as explained in Chapter 5, there are other benefits, such as the travel cost savings benefit, that are enjoyed by both users and non-users. Similarly, in this simulation, we have considered the cost of purchasing a VICS unit that has not only the VICS function but also a car navigation function. However, in order to discuss the optimal penetration rate in relation to this cost, we should account for the benefit of the car navigation function as well. In addition, apart from the benefits to drivers, there are benefits to society as a whole, such as the benefit of reduced emissions of pollutants such as PM, NO_x and CO_2. In the future, we need to refine our measurements to consider these points. In particular, to assess the benefit of car navigation systems, we will need to carry out a survey of drivers to measure the psychological value of such systems by the Contingent Valuation Method (CVM).

Notes

1. 'DID (Densely Inhabited Districts)' and 'other urban areas', in 'Road Traffic Census' (2005) Road Bureau, Japanese Ministry of Construction (now the Japanese Ministry of Land, Infrastructure and Transport).

2. *Jouyousha shijo doukou chosa* (*Passenger car Market Trends*) (2007) Japan Automobile Manufacturers Association, Inc., available at http://www.jama.or.jp/lib/invest_analysis/index.html (accessed November 29, 2010) (in Japanese).
3. Calculation based on 'Road Traffic Census' (2005) Road Bureau, Japanese Ministry of Construction (in Japanese).
4. 'Road Traffic Census' (2005) Road Bureau, Japanese Ministry of Construction (in Japanese).
5. 'Yearly Average of Monthly Receipts and Disbursements per Household in Fiscal Year – All Japan', in *Annual Report on Family Income and Expenditure Survey, 2008*, Japanese Ministry of Internal Affairs and Communications, available at http://www.stat.go.jp/data/sav/2008np/index.htm (accessed November 29, 2010) (in Japanese).
6. Calculated from the average monthly working hours in establishments with five or more employees, in *Annual Report on the Monthly Labor Survey, 2008*, Japanese Ministry of Health, Labour and Welfare (in Japanese).
7. *Kogata kei truck shijo doukou chosa* (*Small-sized Truck and Mini-sized Trucks Market Trends* (2008) and *Futsu truck shijo doukou chosa* (*Ordinary-sized Truck Market Trends* (2008) Japan Automobile Manufacturers Association, Inc., available at http://www.jama.or.jp/lib/invest_analysis/index.html (accessed November 29, 2010) (in Japanese).
8. Calculation based on 'Road Traffic Census' (2005) Road Bureau, Japanese Ministry of Construction (in Japanese).
9. 'Cost–Benefit Analysis Manual' (2008) Road Bureau, City and Regional Development Bureau, Japanese Ministry of Land, Infrastructure and Transport, available at http://www.mlit.go.jp/road/ir/hyouka/plcy/kijun/bin-ekiH20_11.pdf (accessed November 29, 2010) (in Japanese).
10. As will be evident from formulae to be described later, the number of vehicles does not affect the marginal value of the benefit.
11. At this penetration rate, the penetration rates for each type of vehicle are as follows: 24.3 per cent for passenger cars, 28.4 per cent for mini-sized trucks, 51.1 per cent for small-sized trucks, and 100 per cent for ordinary-sized trucks. The ratio of the total benefit for all users (the quantity 'Benefit for users that is reflected in the market' in Figure 5.4) to the social benefit is 80.2 per cent, while benefits enjoyed by all vehicles (the sum of the quantities 'Benefit for users that is not reflected in the market' and 'Benefit for non-users' in Figure 5.4) account for 19.9 per cent.
12. These values of α have been set based on the results of previous studies on the penetration rate of traffic information and the rate of travel time reduction.

References

Emmerink, R. H. M., Verhoef, E. T., Nijkamp, P., and Rietveld, P. (1996) 'Endogenising Demand for Information in Road Transport', *Annals of Regional Science* 30(2): 201–22.

Emmerink, R. H. M., Verhoef, E. T., Nijkamp, P., and Rietveld, P. (1998) 'Information Effects in Transport with Stochastic Capacity and Uncertainty Costs', *International Economic Review* 39(1): 89–109.

Huang, H. J., Liu, T. L., and Yang, H. (2008) 'Modeling the Evolutions of Day-to-day Route Choice and Year-to-year ATIS Adoption with Stochastic User Equilibrium', *Journal of Advanced Transportation*, 42(2): 111–27.

Japanese Automobile Inspection & Registration Association Website available at http://www.airia.or.jp/number/index.html, (accessed November 29, 2010), (in Japanese).

Li, S. G. (2004) 'The Determination of the Endogenous Market Penetration and Compliance Rate of Advanced Traveler Information Systems', *Transport* 19(4): 162–70.

Li, Z. C., Huang, H. J., and Xiong, Y. (2003) 'Mixed Equilibrium Behavior and Market Penetration with Globa Demand Elasticity under Advanced Traveler Information Systems', in *Proceedings of the IEEE 6th International Conference on Intelligent Transportation Systems*, pp. 506–9, New York: IEEE.

Yang, H. (1998) 'Multiple Equilibrium Behavior and Advanced Traveler Information Systems with Endogenous Market Penetration', *Transportation Research Part B* 32(2): 205–18.

Yang, H. (1999) 'Evaluating the Benefit of a Combined Route Guidance and Road Pricing System in a Network with Recurrent Congestion', *Transportation* 26(3): 299–322.

Yang, H. and Meng, Q. (2001) 'Modeling User Adoption of Advanced Traveler Information Systems: Dynamic Evolution and Stationary Equilibrium', *Transportation Research Part A* 35(10): 895–912.

Yin, Y. and Yang, H. (2003) 'Simultaneous Determination of the Equilibrium Market Penetration and Compliance Rate of Advanced Traveler Information Systems', *Transportation Research Part A* 37(2): 165–81.

Zang, R. and Verhoef, E. T. (2006) 'A Monopolistic Market for Advanced Traveler Information Systems and Road Use Efficiency', *Transportation Research Part A* 40(5): 424–43.

7
Market Penetration of Safety-Related ITSs

Masanobu Kii and Hiroaki Miyoshi

7.1 Introduction

One of the most active areas of research on intelligent transport systems (ITSs) is the study of how these systems improve vehicle safety and prevent traffic accidents. Although a number of safety-related ITS technologies have been developed, two particular technologies – the *Roadside Information-Based Driving Support Systems* (hereafter referred as to car-to-infrastructure systems) and the *Inter-Vehicle Communication Type Driving Support Systems* (hereafter referred as to car-to-car systems) – have attracted much recent attention. These systems differ from the autonomous detection-type driving support systems – such as automated following-distance control systems – in that they assist drivers in negotiating situations that cannot be easily recognized or cannot be visually recognized at all. A survey of Japanese traffic accident and traffic fatality statistics[1] for fiscal year 2007 reveals that fully 86 per cent of all accidents in that year were altercations between vehicles, including rear-end collisions, right-angle collisions at intersections, and accidents occurring during left turns.[2] Autonomous detection-type driving support systems offer insufficient promise of preventing such accidents, and for this reason the spread of car-to-car and car-to-infrastructure system technology is eagerly anticipated as an effective mechanism for significantly reducing the frequency of traffic accidents.

Although both car-to-car and car-to-infrastructure systems seek to prevent accidents by providing information to drivers, the economics of the two systems, as characterized by their market penetration processes, are quite different, as described in Chapter 5. On the other hand, traffic density, which depends on urban structure, also has a significant impact on the nature and extent of vehicular technology penetration. These factors, in turn, affect the nature and extent of ITS externalities.

Several previous studies have proposed analytical techniques for assessing the benefit, and the return on investment, of safety-related ITS technologies (see, for example, He *et al.*, 2010; Kulmala, 2010; Leviäkangas and Lähesmaa, 2002; Spyropoulou, 2008; Vaa *et al.*, 2007). However, to our knowledge, the present chapter is the first study to consider the economic features of safety-related ITS technologies – as discussed at length in Chapter 5 – in the context of an investigation of the market diffusion mechanisms of these technologies, the market penetration rates that optimize social benefit, and the government policies that are most effective in achieving these optimal penetration rates.

The purpose of this study is to clarify the economics of safety-related ITS technologies in the context of their urban settings. We begin by summarizing the relevant features of the two classes of ITS technologies and stating some assumptions that simplify our subsequent analyses. We next introduce a simple model of an urban structure and derive a set of traffic conditions based on this model. We then discuss how safety-related ITS on-board units benefit drivers – both drivers who use the on-board units and those who do not – and calculate the expected market penetration rates, as well as the optimal penetration rates, for on-board units. Finally, we analyze the sensitivity of the economic benefits of safety-related ITS technologies to urban structure.

7.2 The differences between car-to-infrastructure and car-to-car technologies – and some simplifying assumptions

In this section we discuss the key differences between the two types of system introduced above, focusing in particular on two perspectives: (1) accident prevention performance, and (2) economic properties. This discussion then sets the stage for a series of assumptions we will make to simplify the analyses presented in the remainder of this chapter.

7.2.1 Differences in accident prevention performance

We begin by noting that, from an accident prevention perspective, car-to-car and car-to-infrastructure systems do not offer identical opportunities for accident prevention, nor do the two types of system offer the same potential to provide services to users. According to Japan's Ministry of Internal Affairs and Communications (MIC) (2009), there are five types of accidents that both systems can help to prevent: (1) right-angle collisions at intersections, (2) rear-end collisions, (3) left-turn collisions, (4) right-turn undercut collisions,[3] and (5) lane-change collisions. Next, there is one type of collision that car-to-car system technology alone can help to prevent: (6) frontal collisions. There is also one type of information service that car-to-car system technology alone can provide, namely (7) ambulance information. On the other hand, there are two types of accidents that car-to-infrastructure system technology alone can help to prevent: (8) collisions during merging, and (9) collisions involving pedestrians. Finally, there are three types of information service that car-to-infrastructure system technology alone can provide: (10) traffic signal information, (11) information on traffic restrictions, and (12) information on roads.

7.2.2 Differences in economic properties

Let us now briefly review the discussion of Chapter 5 regarding the different economics of car-to-car and car-to-infrastructure systems.

Car-to-car systems are designed to avoid accidents through inter-vehicle communication. An advantage of such a system is that it can be used everywhere, as long as all participating vehicles are equipped with the on-board unit; there is no need to install a roadside infrastructure. On the other hand, if either vehicle in a potential altercation is *not* equipped with the on-board unit, then the vehicles cannot communicate with each other to avert an accident. Thus, this system can work only if all participating vehicles are equipped with the on-board unit; the more system users, the greater the benefit to each user. This is known as *network externality* (Katz and Shapiro, 1985, 1994; Libenstein, 1950; Rohlfs, 1974). The downside of this fact is that, as long as the number of vehicles equipped with car-to-car system on-board units remains low, consumers have no incentive to purchase the on-board units, and the on-board units will not diffuse throughout the marketplace.

On the other hand, car-to-infrastructure systems are based on information collected from vehicle detectors installed on roads. Thus, these systems work only in locations where roadside devices are installed. This type of system works as long as *any* of the relevant vehicles have the on-board unit. Thus, the benefit to system users is independent of the number of system users. In addition, car-to-infrastructure systems, unlike car-to-car systems, offer benefits not only to users but also to *non*-users, who experience fewer accidents thanks to evasive actions taken by users to avoid collisions. The downside of this fact is that, as on-board units become more and more widespread throughout society, more and more free-riders will be able to benefit in the case where both vehicles in a potential altercation are able to receive warning information; as a result, the market value to the consumer of the car-to-infrastructure system on-board unit diminishes, and this can retard the spread of car-to-infrastructure system on-board units.

Japan's MIC (2009) has proposed a *dual mode* strategy, in which both types of system are developed and deployed, with the rationale that (1) the simultaneous existence of mechanisms for both vehicle-to-vehicle and road-to-vehicle information exchange will increase opportunities for providing valuable services to users, and (2) if, eventually, a single board unit could be used for both vehicle-to-vehicle and road-to-vehicle communication, the corresponding simplification in system architecture would improve the cost performance and other aspects of the system. In this chapter, we will investigate this *dual mode* proposal using estimated market price–demand curves; our analysis will lead us to argue in favor of a *dual mode* system, but for reasons different from those invoked by MIC.

7.2.3 Simplifying assumptions for simulation analysis

We now introduce three assumptions, based on our preceding discussion of the differences between car-to-car and car-to-infrastructure system technologies, that we will make to simplify the simulation analyses presented in the remainder of this chapter.

First, we assume that there is no distinction between the types of accidents that may be avoided through use of car-to-car and car-to-infrastructure system technologies. In reality, as discussed above, there are some differences in the types of accidents that the two types of system can help avoid. However, for simplicity in this study we assume that both types of system are effective in preventing *all* types of accidents that arise when vehicles pass one another.

Second, we assume that the use of either a car-to-car *or* a car-to-infrastructure system will suffice to prevent *all* traffic accidents that a system user might otherwise encounter. In reality, of course, the actual accident prevention rate will vary significantly depending on a large number of factors (such as, for example, whether the system merely passively alerts the driver to the presence of a road hazard or actively assumes control of the vehicle in an attempt to evade the hazard). However, for simplicity in this study we consider idealized models of car-to-car and car-to-infrastructure systems that are capable of preventing 100 per cent of all collisions.

Third, we assume that both cars involved in a potential accident receive the warnings and other information needed to avoid the accident. The systems currently under consideration in Japan are designed in such a way as to deliver warnings to only *one* of the two parties in a potential accident – namely, whichever party is potentially guilty of a more serious infraction, or is likely to sustain more serious damage. For example, in the event of a left-turn collision, only the left-turning vehicle would receive warning notifications. However, such a scheme would be atypical among the various systems currently under consideration around the world, and even in Japan some manufacturers have adopted systems that deliver warnings to both vehicles. As discussed in Chapter 5, the market values of car-to-infrastructure systems differ widely depending on whether the system warns only one vehicle or both vehicles of impending collisions. When the system warns only one vehicle, there is no mechanism for free-riders to benefit from the system, and hence no corresponding degradation in the market value of the system. Nonetheless, for simplicity in this study we assume that car-to-infrastructure systems always deliver warnings to both vehicles in a potential collision.

Our forth assumption is that the social cost of safety-related ITS is the sum of the cost of roadside devices and the cost of supplying on-board units, and that the former is a fixed cost (an initial cost), while the latter is a variable cost that varies with the number of on-board units. In addition, we assume that the production function of on-board units displays constant returns to scale. Under these assumptions, the marginal social cost of an on-board unit is equal to the price of an on-board unit.

7.3 Urban structure and traffic patterns

As discussed above, car-to-car and car-to-infrastructure systems exhibit different economics. The two systems may also exhibit different impacts on accident reduction in a given region, depending on the urban structure. A car-to-car system will be effective if the on-board units are diffused sufficiently thoroughly over the region in question, while a car-to-infrastructure system works only at locations where roadside devices are installed. This means that geographical traffic patterns are important inputs to consider when evaluating the performance of these systems, particularly for the case of car-to-infrastructure systems.

7.3.1 A model urban system

To illustrate the analyses in this study, we consider the simple linear model of an urban system (Anas and Xu, 1999; Fujita and Ogawa, 1982) depicted in Figure 7.1, in which the population density $q(x)$ attains a

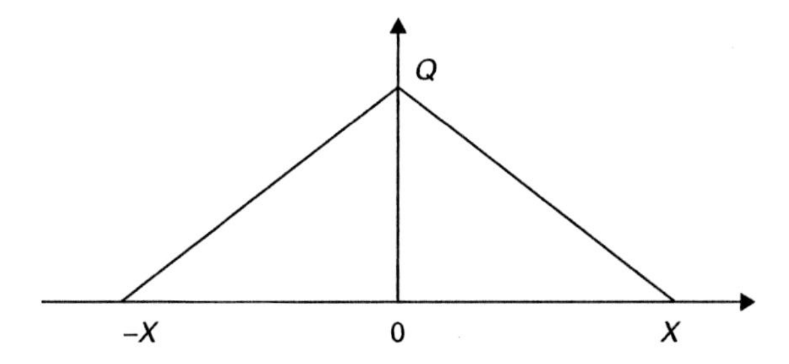

Figure 7.1 Linear model of urban population density

maximum value of Q at the origin and falls linearly to zero at the urban edges, $\pm X$:

$$q(x) = Q\left(1 - |x|/X\right). \tag{7.1}$$

Although a linear model might seem overly simplistic for a discussion of the impact of ITS technology in the real world, this simplified model does afford some insight into the impact of spatial factors on the efficacy of safety-related ITS technologies. In the future, more sophisticated spatial models may be required to capture the full complexity of traffic patterns within a particular urban structure. However, for the purposes of this study, the simple model considered here already affords sufficient perspective to suggest directions for future simulation studies.

7.3.2 Traffic pattern assumptions

Each commuter is assumed to make one round trip per day between an origin x and a destination y determined solely by the population density. More specifically, an individual commuter departs from an origin point in the range $[x, x+dx]$ with probability $Pr(x)\,dx$, and travels to a destination point in the range $[y, y+dy]$ with probability $Pr(y)\,dy$, where the function Pr is defined by

$$\Pr(x) = \left(1 - |x|/X\right)/X. \tag{7.2}$$

Departure times are assumed to be distributed uniformly, so that the probability that any one commuter departs between time t and time $t+dt$ is simply dt/T (independent of t) where $T = 24$ hours (one full day). All commuters are assumed to travel at a constant speed v irrespective of origin, destination, and departure time.

We assume that an accident is possible whenever vehicles pass each other traveling in opposite directions. Consider two vehicles, one starting

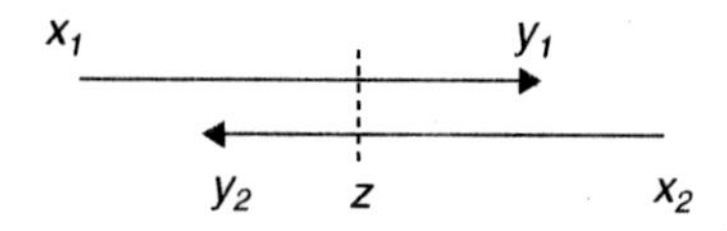

Figure 7.2 Meeting point of two vehicle trajectories

from location x_1 at time t_1 and the other from x_2 at t_2 ($x_1<x_2$). These two vehicles meet at a point z (see Figure 7.2) given by the expression

$$z = \frac{-t_1 + t_2}{2} v + \frac{x_1 + x_2}{2}. \tag{7.3}$$

If the destinations of these vehicles are y_1 and y_2, respectively, then the condition that the two vehicles pass each other is $x_1<z<y_1$ **and** $x_2>z>y_2$.

We can now derive an expression for the rate of collisions occurring between two spatial points z and $z+dz$. Consider first a single vehicle departing from location x_0 at time t_0 and traveling rightward. The number of left-moving vehicles that this vehicle passes in the spatial interval $[z, z+dz]$ is

$$n_L(z)dz = \frac{dz}{vT} \cdot \int_{-X}^{z} q(y)dy \cdot \int_{z}^{X} \Pr(y')dy'. \tag{7.4}$$

This expression is independent of $x0$ and $t0$; thus *any* right-moving vehicle, independent of origin or departure time, encounters precisely $n_L(z)\,dz$ left-moving vehicles as it travels between z and $z+dz$. Summing over all origins x_0, the total number of right-moving vehicles arriving at z in a time interval dt is

$$n_R(z)dt = \frac{2dt}{T} \cdot \int_{z}^{X} q(x)dx \cdot \int_{-X}^{z} \Pr(x')dx'. \tag{7.5}$$

(The factor of 2 in Eq. (7.5) accounts for the fact that each commuter makes two trips, one trip to work and a second trip home, in each time interval T). Then the number of head-on crossings in the spatial interval $[z, z+dz]$ and in a time interval of length dt is

$$n(z)dtdz = N_R(z)dt \cdot N_L(z)dz.$$

$$=\left(\frac{2dzdt}{vT}\right)\cdot\int_{-X}^{z}q(y)dy\cdot\int_{z}^{X}\Pr(y')dy'\cdot\int_{z}^{X}q(x)dx\cdot\int_{-X}^{z}\Pr(x')dx'. \tag{7.6}$$

Inserting (1) and (2) into (6), the density of crossings at z is

$$n(z)=\frac{Q^2}{8vT^2\cdot X^6}\cdot(X-|z|)^4\cdot\left(2X^2-(X-|z|)^2\right)^2. \tag{7.7}$$

Figure 7.3 plots the density of head-on crossings for a city of radius 20 km and population 1 million, assuming a commute speed of $v = 20$ km/h.

The total distance traveled by all commuters, and the total number of crossings across the city per day, are important quantities. In our model these quantities take the values

$$D=4\cdot\int_{-X}^{X}q(x)\cdot\int_{x}^{X}\Pr(y)\cdot(y-x)dydx=\frac{28}{30}\cdot Q\cdot X^2, \tag{7.8}$$

$$N_a=2T\int_{0}^{X}N(z)dz=\frac{107}{1260}\cdot\frac{Q^2\cdot X^3}{vT}. \tag{7.9}$$

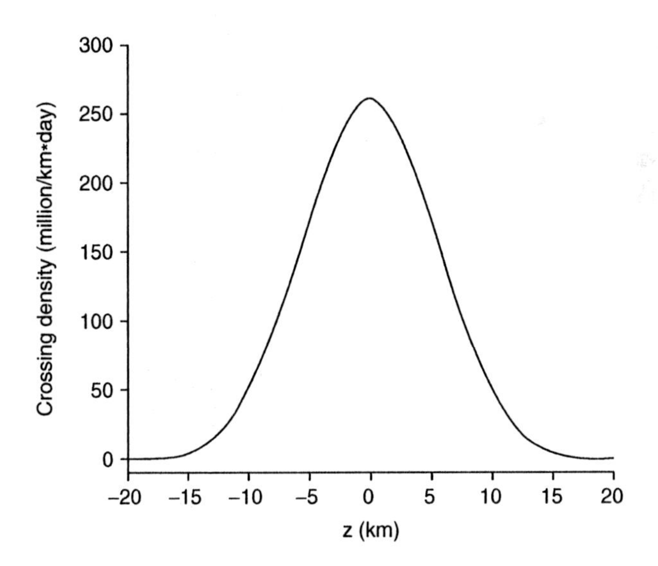

Figure 7.3 Spatial distribution of head-on crossings

7.3.3 Market penetration rates and accident reduction probabilities

We denote the penetration rate of the safety-related ITS on-board unit by α and assume that the on-board units are uniformly distributed throughout the urban space. In other words, the fraction of vehicles equipped with the on-board unit is α everywhere throughout the city. Given these assumptions, when two vehicles cross head-on, the probability that both vehicles have the on-board unit is α^2, and the probability that either one or both vehicles have the on-board unit is $2\alpha - \alpha^2$.

As discussed above, car-to-car system technology prevents accidents only when both vehicles are equipped with the on-board unit. For this type of system, the overall accident reduction rate is thus α^2 over the full urban area. A vehicle equipped with the on-board unit will avoid accidents with probability α (the probability that the head-on-crossing vehicle has the on-board unit), while a vehicle without the on-board unit has zero probability of avoiding accidents.

Car-to-infrastructure system technology, on the other hand, prevents accidents when at least one vehicle is equipped with the on-board unit. Thus the accident reduction rate for this type of system is $2\alpha - \alpha^2$ over the coverage area of the roadside communicator. A vehicle with the on-board unit avoids accidents with probability 1, while a vehicle without the on-board unit – in contrast to the case of the car-to-car system – also enjoys a non-zero accident reduction probability of α (the probability that the head-on crossing vehicle has the on-board unit).

Table 7.1 Accident-avoidance performance of car-to-car and car-to-infrastructure system technologies

Accident-avoidance rate	Car-to-car	Car-to-infrastructure
For users	α	1
For non-users	0	α
Total	α^2	$2\alpha-\alpha^2$
Effective area	All over the city	Roadside communicator installed

Note: α = penetration rate of the safety-related ITS on-board unit.

Table 7.1 summarizes the accident-avoidance performance of the two systems. These qualitative features can be quantified by using the on-board unit penetration rate to define accident-avoidance probabilities. The network externality of the car-to-car system is represented in this study by the accident-avoidance probability of α experienced by on-board unit users; as the number of users increases, the benefit derived by each user increases. The car-to-infrastructure system, in contrast, does not have this feature; users of this system enjoy a constant accident-avoidance probability of 1 independent of the number of system users. As for non-users, the externality to non-users of the car-to-infrastructure system is represented by the accident-avoidance probability of α enjoyed by non-users; this demonstrates that, as the number of system users increases, the benefit to *non-users* increases with it. On the other hand, with the car-to-car system, non-users experience an accident-avoidance probability of zero independent of the number of system users. These differences between the car-to-car and car-to-infrastructure systems will affect the market penetration rates of the two system on-board units, and may require different policies for promoting traffic safety through the use of ITS technologies.

7.4 Estimated and optimal market penetration rates for safety-related ITS technologies

To illustrate the different impacts exhibited by the two types of system, in this section we discuss the results of numerical experiments performed to simulate the ITS market. In addition to the assumptions discussed above, we assume the urban population characteristics and accident frequency rates tabulated in Table 7.2. The accident frequency rates in this table are derived from Japanese road safety statistics.

We assume that the lifetime of on-board unit ITS device is 10 years[4] and that the discount rate for on-board units is 7 per cent. We also assume that value lost in a traffic accident varies according to a log-normal distribution. Figure 7.4 plots the probability density distribution for total value lost in fatal traffic accidents. The average value lost in a fatal traffic accident is 246 million yen per death, based on a report on the economic valuation of traffic accidents by Japan's Ministry of Land, Infrastructure, Transport and Tourism (2008); this report does not give the distribution of value lost, and we arbitrarily set the standard deviation to one-half the average value in this study. The probability density distribution for total value lost due to injuries sustained in traffic accidents is derived on the basis of similar assumptions; the average value

Table 7.2 Urban population characteristics and accident rates

Urban population	1 million
Population density at city center	50,000/km
Urban radius	20 km
Commute speed	20 km/h
Total daily travel distance (Eq. 7.8)	19 million km/day
Number of head-on crossings (Eq. 7.9)	7.1 billion/day
Accident frequency (derived from Japanese road safety statistics)	
Fatal	2.8 people / billion km
Injury (severe)	35 people / billion km
Injury (slight)	727 people / billion km
Accident rate per crossing (estimated based on calculated total travel distance, number of head-on crossings, and the accident frequency)	
Fatal	1.4×10^{-11}
Injury (severe)	1.6×10^{-10}
Injury (slight)	2.5×10^{-9}

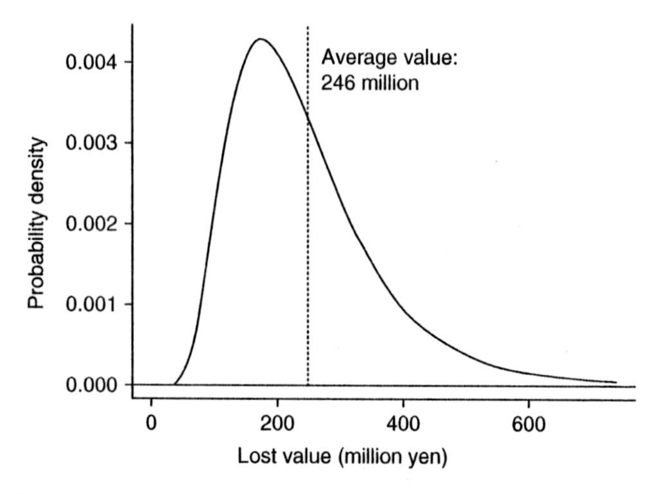

Figure 7.4 Probability density distribution for value lost in fatal traffic accidents

lost due to severe and slight injuries sustained in traffic accidents is 9.26 million yen and 1.38 million yen respectively. On-board units are more densely diffused among people with higher potential lost value – the higher the potential for loss due to a traffic accident, the greater the benefit from avoiding an accident.

7.4.1 Car-to-car system on-board units

Assuming a penetration rate of α, the maximum benefit available to a potential new user of a car-to-car system on-board unit is

$$ED_c = \alpha B \, \Gamma \, P_a, \text{ where } \alpha = \int_B^{\infty} \Pr_L(b) db \ . \qquad (7.10)$$

Here ED_c is the market value to a new user over the lifetime of the on-board unit – that is, the expected equilibrium demand curve; B is the minimum value lost, at a penetration rate of α, by system users due to traffic accidents; α is a multiplicative factor that determines the discounted total benefit over the lifetime of the on-board unit; P_a is the expected accident rate; and $\Pr_L(b)$ is the probability density for lost value b. Due to the continuity of the lost-value distribution, the benefit B is also equal to the maximum lost value among non-users. B is computed under the conditions discussed above and substituted into Eq. (7.10) to calculate ED_c.

We next determine the social benefit (SB_c) – which represents the total benefit to society derived from diffusion of the car-to-car system on-board unit throughout the marketplace – as well as the marginal social benefit (MSB_c), which represents the additional benefit to society derived from the addition of one user of the car-to-car system on-board unit. These quantities are obtained as follows:

$$SB_C = n \cdot \alpha \cdot \Gamma \cdot P_a \int_B^{\infty} \Pr_L(b) db, \qquad (7.11)$$

$$MSB_C = \frac{1}{n} \frac{\partial SB}{\partial a} = \left(\int_B^{\infty} b \cdot \Pr_L(b) db + \alpha \cdot B \right) \Gamma \cdot P_a \, . \qquad (7.12)$$

Here n is the total number of cars present in society.

Based on these assumptions, Figure 7.5 plots ED_c and MSB_c as a function of the market penetration rate, with the horizontal dashed line representing an assumed on-board unit price of 20,000 yen. ED_c peaks at an intermediate penetration rate, beyond which the benefit rapidly drops as the penetration rate increases. Higher penetration rates mean that the on-board unit is installed by drivers with lower potential lost value. Thus the marginal user's benefit drops at high penetration rates.

On the other hand, the value of MSB_c exceeds the value of ED_c at all values of the market penetration rate. This represents the fact that, whenever one additional user joins the system, all existing system users

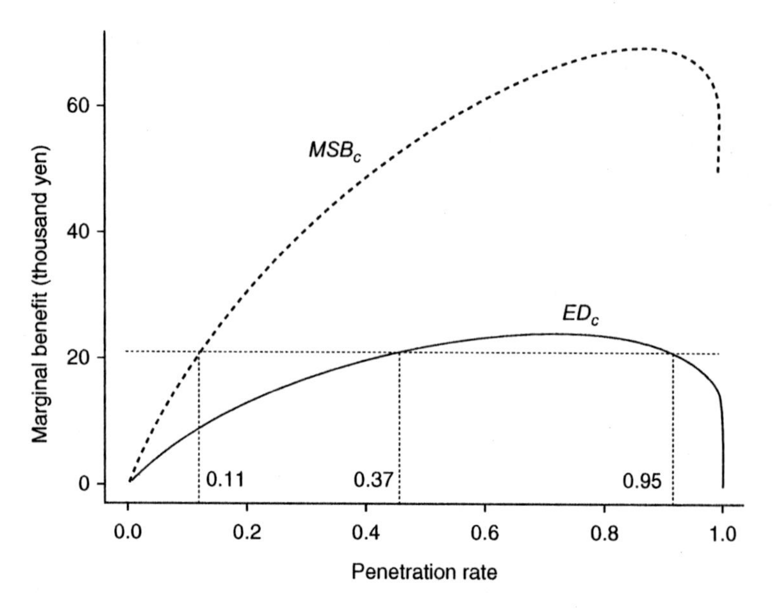

Figure 7.5 Market penetration rate and optimal penetration rate (car-to-car system)

benefit from the corresponding decrease in the overall accident rate. The difference between MSB_c and ED_c shows the magnitude of network externality.

The expected market penetration rate is determined by the intersection of the price curve with the equilibrium demand curve (*ED*c). Figure 7.5 indicates that these two curves intersect at two distinct market penetration rates, namely, 37 per cent and 95 per cent. The former of these two values corresponds to the penetration rate we termed the 'critical mass' in Chapter 5. If the market penetration rate falls below this value, the market will inevitably contract, whereas once the market penetration rate reaches this value it will automatically continue to increase until it reaches a value of 95 per cent. The fact that the critical mass in this case is predicted to be the abnormally large value of 37 per cent reflects the difficulties inherent in allowing the car-to-car system on-board unit to diffuse throughout the marketplace based on free-market mechanisms alone.

On the other hand, the optimal market penetration rate is determined by the intersection of the price curve with the marginal social benefit curve (*MSB*c), as depicted in Figure 7.5. The *MSB* curve and price line intersect at a market penetration rate of 11 per cent. However, although this value does correspond to a market penetration rate at which the marginal social benefit equals the price, this is not the point at which total social benefit is maximized. Instead, as discussed in Chapter 5, the optimal market penetration rate in this case is 100 per cent. This optimal market penetration rate cannot be achieved by free-market mechanisms alone, and instead we must rely on policies such as government subsidies – either to consumers or to manufacturers – to offset the difference in value between *MSB*c and *ED*c at the optimal penetration rate (which, in this figure, is the full cost of the board unit), thus promoting the board units to the status of standard equipment in new vehicles.

7.4.2 Car-to-infrastructure system on-board units

Assuming a penetration rate of α and an upper bound of R_c on the rate of head-on crossings near roadside communicator devices installations, the benefit, BR_u, enjoyed by a new user of a car-to-infrastructure system on-board unit is given by

$$BR_u = R_c\, B\, \Gamma\, P_a, \text{ where } \alpha = \int_B^{\infty} \Pr_L(b)\,db \tag{7.13}$$

With the car-to-infrastructure system, the benefit enjoyed by users does not depend on the penetration rate, because the accident-avoidance probability for users is unity independent of the number of other users. Therefore, the benefit enjoyed by a user depends only on the potential value lost by that user.

Under this system, users avoid accidents whether or not the head-on-crossing vehicle is equipped with the on-board unit. Thus non-users also benefit from the system. The maximum benefit to non-users, BR_o, is given by

$$BR_o = \alpha\, R_c\, B\, \Gamma\, P_a. \tag{7.14}$$

The market value of the board unit for a marginal user must be thought of as the difference between the traffic accident prevention benefit realized by a marginal user and that realized by a non-user, as described in Chapter 5. The market value of the on-board system ED_r – that is, the expected equilibrium demand curve – is the difference between the benefits to users and to non-users:

$$ED_r = BR_u - BR_o = (1-\alpha)\, R_c \cdot B \cdot \Gamma \cdot P_a. \tag{7.15}$$

We next consider the social benefit (SB_r). Because car-to-infrastructure system technology tends to benefit non-users as well as users, we have

$$SB_r = n \cdot R_C \cdot \Gamma \cdot P_a \int_B^{\infty} b \cdot \Pr_L(b)db$$
$$+ n \cdot \alpha \cdot R_C \cdot \Gamma \cdot P_a \int_0^B b \cdot \Pr_L(b)db. \tag{7.16}$$

The first and second terms on the right-hand side of Eq. (7.16) are the benefits realized by system users and by non-users, respectively.

Next, differentiating Eq. (7.16) yields an equation for the marginal social benefit (the increase in social benefit arising from a unit increase in penetration rate):

$$MSB_r = \frac{1}{n}\frac{\partial SB}{\partial \alpha} = R_C \cdot \Gamma \cdot P_a \left(\int_0^B b \cdot \Pr_L \cdot \Pr_L(b)db + (1-\alpha)B \right). \tag{7.17}$$

Figure 7.6 plots the benefits enjoyed by users and by non-users, as well as the expected equilibrium demand curve, as a function of the penetration

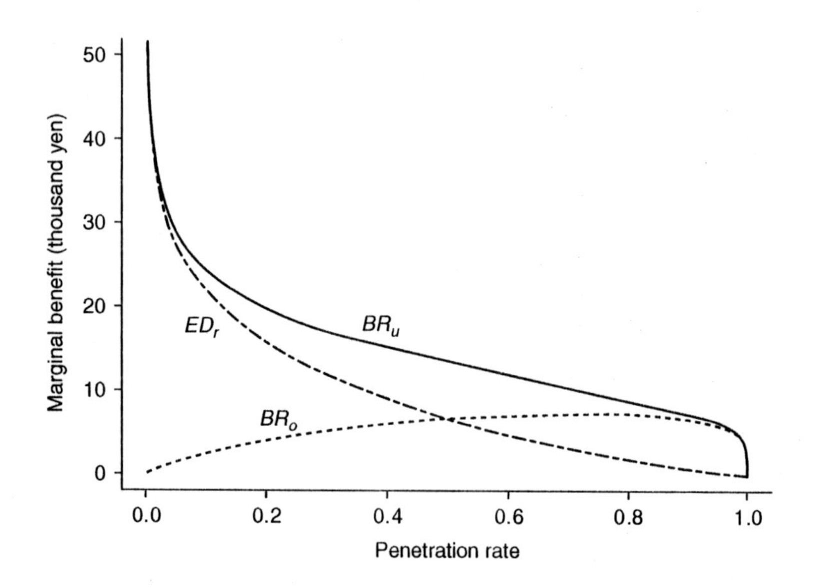

Figure 7.6 Benefit to users, benefit to non-users, and expected equilibrium demand curve as a function of on-board unit penetration rate α (for the car-to-infrastructure system)

rate α. Here we have assumed that roadside communicators are installed at points ±2 km from the center. The benefit to users decreases as the penetration rate increases, because the marginal user's lost value due to an accident decreases as the penetration rate increases. The benefit to non-users increases at first, because the accident-avoidance probability increases, but eventually peaks and begins to decline due to the decrease in marginal lost value. The market value of the system, defined as the difference between these benefits, declines as the penetration rate increases.

Figure 7.7 plots curves ED_r and MSB_r, as well as the expected and optimal market penetration rates assuming a price of 20,000 yen, for the car-to-infrastructure system. We note that MSB_r exceeds ED_r at all values of the penetration rate. The difference between MSB_r and ED_r represents a social benefit not reflected by the market price; in other words, this quantity is the marginal value of the sum of the quantities termed 'benefit for non-users' and 'benefit for users that is *not* reflected in the market' in Chapter 5.

The market penetration rate determined by the intersection of curve ED_r and the price curve is 12 per cent. Because curve ED_r is rapidly decreasing, the market equilibrium occurs at an extremely low market penetration rate. On the other hand, the optimal market penetration

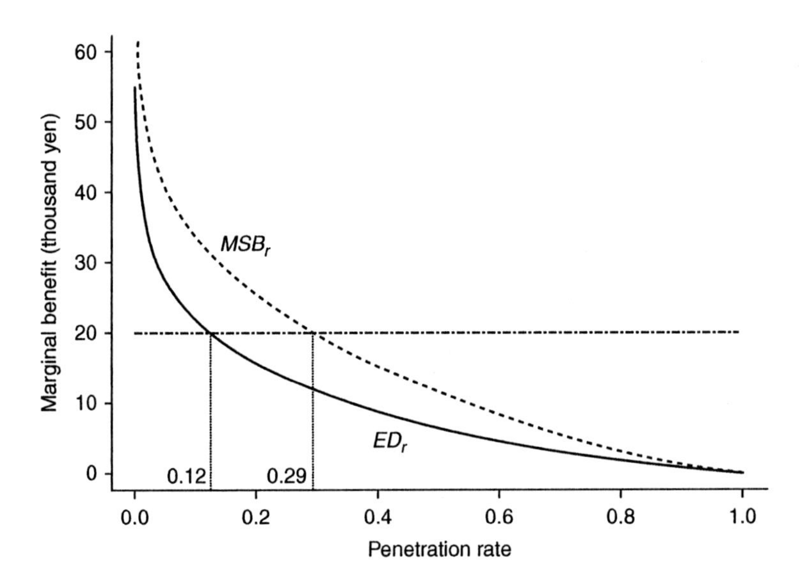

Figure 7.7 Market penetration rate and optimal penetration rate (car-to-infrastructure system)

rate, determined by the intersection of curve MSB_r with the price curve, is 29 per cent. In this case, as in the case of the car-to-car system, market mechanisms alone are insufficient to achieve the optimal market penetration rate, and instead government policies, in the form of subsidies to users to offset a portion of the cost of the board unit, are required.

7.4.3 Dual-mode system on-board units

In the case of the car-to-infrastructure system, the market penetration rate determined by free-market mechanisms is extremely low. Moreover, in the case of the car-to-car system, the board units will not diffuse through the market unless the critical mass is met or exceeded. These considerations lead us to introduce the concept of a *dual-mode* on-board unit, which works as a car-to-infrastructure system on-board unit in areas where roadside communicators are installed and as a car-to-car system on-board unit in other areas. On the basis of the assumptions discussed above, the market value to a new user of the on-board unit – that is, the expected equilibrium demand curve ED_d – is given by Eqs (7.10) and (7.15) as

$$ED_d = ED_r + (1 - R_c)\, ED_c. \tag{7.18}$$

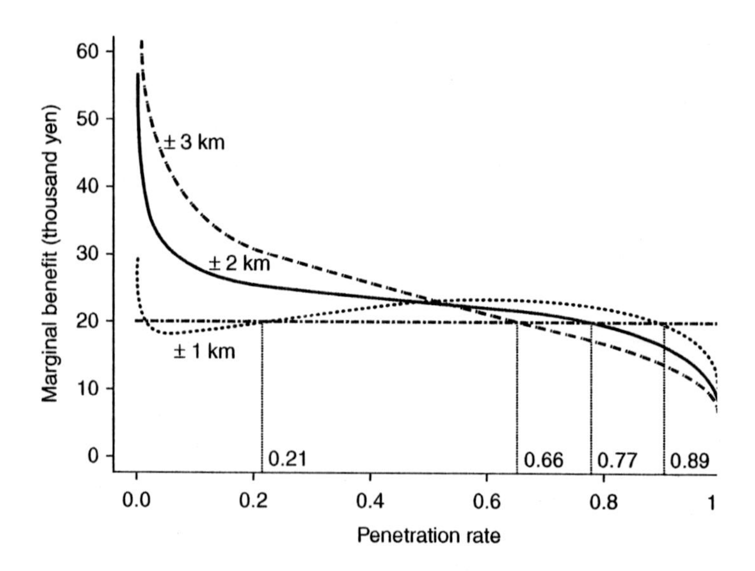

Figure 7.8 Expected equilibrium demand curves for dual-mode on-board units, assuming three different values of the roadside communicator range

In this case, the social benefit and the marginal social benefit are given, respectively, by Eqs (7.11) and (7.16) and by Eqs (7.12) and (7.17) as

$$SB_d = SB_r + (1 - R_c)\, SB_c, \tag{7.19}$$
$$MSB_d = MSB_r + (1 - R_c)\, MSB_c. \tag{7.20}$$

Figure 7.8 plots the expected equilibrium demand curve ED_d, as a function of the on-board unit penetration rate α, for roadside communicators installed at ±1 km, ±2 km, and ±3 km from the urban center. For the ±2 km case, the difference between benefit and cost is positive at zero penetration rate, and the rate naturally increases to 77 per cent. For the ±3 km case, the benefit is initially higher than in the ±2 km case, but declines steeply thereafter and becomes equal to the cost at a penetration rate of 66 per cent. For the ±1 km case, the initial benefit is below cost, but the system will diffuse naturally once the penetration rate exceeds a critical mass of 21 per cent, eventually increasing to 89 per cent.

The market value of a dual-mode on-board unit in regions in which roadside communicators are installed is defined in the same way as for a car-to-infrastructure system on-board unit – namely, as the difference between the benefits to users and to non-users. Therefore, with higher penetration rates, the marginal benefit of a dual-mode on-board unit is lower than that of a car-to-car system on-board unit. Consequently, the larger the roadside device installation area, the higher is the average benefit to non-users. Therefore, the benefit in the ±3 km case declines steeply, and the equilibrium penetration rate is lower than in the ±2 km case. On the other hand, if the roadside installation area is too small, the benefit of the car-to-infrastructure system does not exceed the cost, and some kind of policy intervention is required for penetration, as is true for the car-to-car system.

Figure 7.9 plots the marginal social benefit curve MSB_d, as a function of the on-board unit penetration rate αb, for roadside communicators installed at ±1 km, ±2 km, and ±3 km from the urban center.

At low values of the roadside device installation coverage area, the marginal social benefit curve for the dual-mode on-board unit resembles that of the car-to-car on-board unit. However, as the roadside device installation area increases, the marginal social benefit curve for the dual-mode on-board unit increasingly resembles that of the car-to-infrastructure on-board unit. However, in any case, the optimal market penetration rate for the on-board unit is 100 per cent (at least assuming a on-board unit price of 20,000 yen). Note that this analysis has entirely neglected

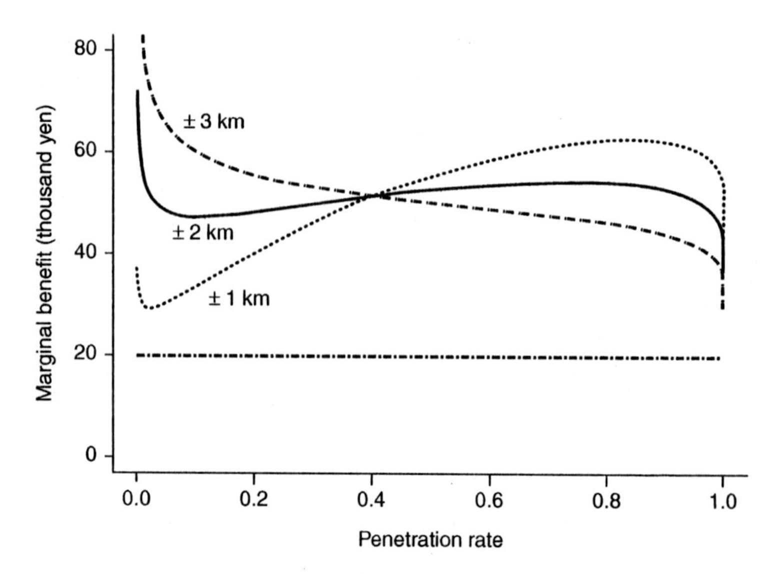

Figure 7.9 Marginal social benefit curves for dual-mode on-board units, assuming three different values of the roadside communicator range

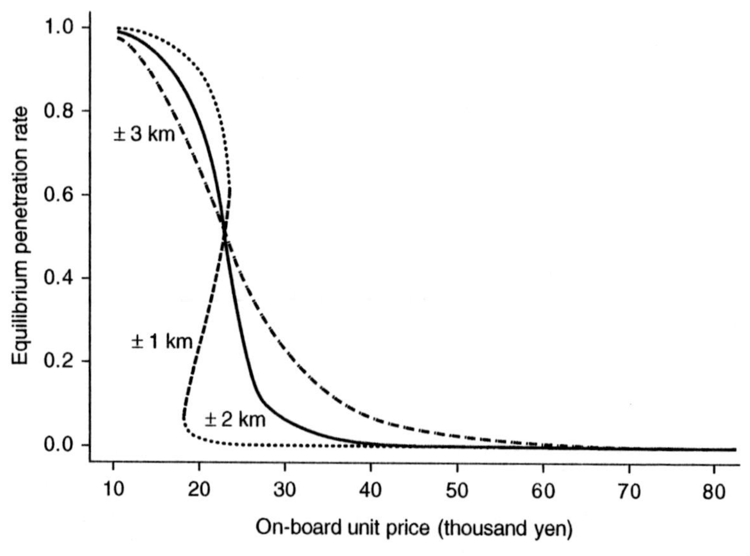

Figure 7.10 On-board unit price versus equilibrium penetration rate

the fixed costs associated with installing the roadside-device infrastructure; whether or not there is a net benefit to society at the optimal market penetration rate is a question requiring further investigation.

We next analyze the impact of on-board unit pricing on the market penetration dynamics of the system. Above we fixed the cost of the board unit at 20,000 yen, but at present no on-board units have been released for sale, and the actual price at which a real-world on-board unit will eventually be sold remains uncertain. Figure 7.10 plots the equilibrium penetration rate as a function of on-board unit price over the range 10,000 – 80,000 yen. This plot indicates that, in the regime in which board units cost 25,000 yen or more, the penetration rate increases as the area covered by roadside communicator installations increases. On the other hand, in the lower-price regime, and with roadside communicators chosen to have a coverage range of ±1 km, we see that multiple equilibria exist; the solid line indicates the stable equilibrium, while the dashed line corresponds to the unstable equilibrium. The unstable equilibrium indicates the existence of a *critical mass* under the given pricing conditions; if the price of the board unit may be expected to fall below a certain threshold (23,000 yen in this study), then an effective strategy for achieving widespread use of the system is to reduce the area covered by the roadside communicators, while offering subsidies or other assistance toward the purchase of the board units, thus stimulating market penetration until the critical mass is attained. On the other hand, if the price of the on-board unit is not expected to fall sufficiently low, then subsidies are an ineffective means of stimulating penetration; in this case, the most effective way to promote more widespread use of the system is to increase the area covered by roadside communicators.

As we see from Figures 7.8 and 7.9, externalities ensure that the optimal penetration rate is always higher than the equilibrium penetration rate. If the price of the on-board unit is less than 47,000 yen in the ±2 km case, the marginal social benefit curve is always above supply curve. It suggests that there are large gaps between the optimal and equilibrium penetration rates at certain range of on-board unit price (in this study, at least over the range between 23,000 and 47,000 yen). This implies that some public policies that intervene in the market to promote the on-board units will lead to a large improvement in social welfare.

To analyze the sensitivity of the dual-mode on-board unit to the urban structure, we considered urban radii of 15, 20, and 25 km, in all cases with the same total population and with roadside communicators installed at points ±2 km from the urban center. Figure 7.11

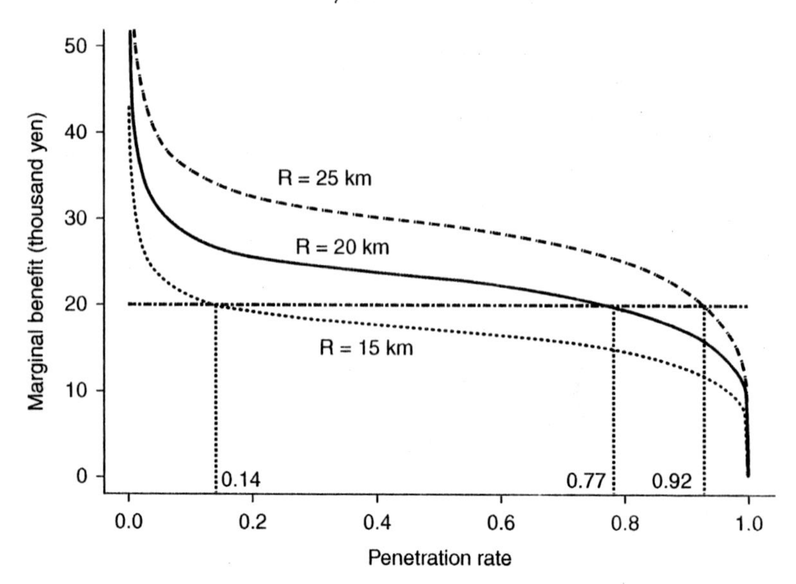

Figure 7.11 Expected equilibrium demand curves for dual-mode on-board units, assuming three different urban structure models

plots the expected equilibrium demand curve in each of the three cases; evidently, the larger the urban area, the greater the benefit. This reflects the fact that daily travel distances increase as the urban radius increases. The equilibrium penetration rate under the condition that on-board unit price is 20,000 yen is just 14 per cent when the urban radius is 15 km, but jumps to 92 per cent for an urban radius of 25 km. This result suggests that urban structure has a significant impact on the penetration of safety-related ITS.

7.5 Conclusions

In this study, we examined the economics of various safety-related ITS on-board units in the context of a model urban structure. Our primary findings are as follows.

First, car-to-infrastructure systems exhibit positive network externality. This causes the market value of the car-to-infrastructure system on-board unit to decline as the on-board unit diffuses through the market. On the other hand, car-to-car systems exhibit negative network externality. The market value of the car-to-car system on-board unit peaks at a particular value of the on-board unit penetration rate. This on-board unit thus requires some form of fiscal support, such as a subsidy, to achieve the critical mass required to attain the optimal market penetration rate.

Second, the dual-mode system is an intermediate entity combining the best features of car-to-infrastructure and car-to-car system technologies. At lower penetration rates, the economics of the dual-mode on-board unit resemble those of the car-to-infrastructure system on-board unit, ensuring that the on-board unit will diffuse naturally throughout the market. At higher penetration rates, the dual-mode on-board unit exhibits the favorable attributes of the car-to-car system on-board unit.

Third, if the area covered by roadside communicator installations is too small, the economic benefit of the car-to-infrastructure system falls short of the critical mass needed to achieve mass adoption of the car-to-car system. On the other hand, if the coverage area is too large, the externality of the car-to-infrastructure system provides enough benefits to non-users to discourage adoption of the on-board unit; in this case the equilibrium penetration rate slides downhill, leading to a decline in overall social welfare.

Fourth, the equilibrium penetration rate depends heavily on the price of the on-board units. If the price is not expected to drop significantly, the most effective way to stimulate market penetration is to increase the area covered by roadside communicator installations. On the other hand, if the price *is* expected to drop significantly, then an effective strategy for promoting market penetration is to reduce the area covered by roadside communicator installations, while offering subsidies or other assistance toward the purchase of the on-board units.

Fifth, the optimal penetration rates of the car-to-car system on-board unit and the dual-mode on-board unit exceed the optimal penetration rate of the car-to-infrastructure system on-board unit. Indeed, both the car-to-car and dual-mode on-board units exhibit an optimal market penetration rate of 100 per cent over a wide range of on-board unit prices in this simulation analysis.

Finally, urban structure has a significant impact on the efficacy of safety-related ITS technologies. This suggests that urban structure issues must be critical inputs to the ITS policymaking process.

As discussed above, the dual-mode system exhibits the favorable behavior of the car-to-infrastructure system at low market penetration rates, allowing the system to attain the critical threshold market penetration rate required for market penetration of the car-to-car system. Once the on-board unit has diffused sufficiently thoroughly through the market, the roadside installation infrastructure may well become useless – or provide quite limited effects. In such cases, future investments in roadside communication infrastructure will yield minimal benefit. However, other advanced systems – systems that depend on coordination with traffic signals, for example – still require roadside devices. Similarly, large traffic intersections may continue to require information transmitters, even if all vehicles are equipped with car-to-car system on-board units. Thus, we do not consider investments in roadside communication infrastructure to be wasteful, although the above analysis reveals the car-to-infrastructure system to be an instance of pump-priming.

In closing, we note that the analyses reported in this chapter employed a simplistic spatial model of urban population density, as well as several theoretical idealizations. In future work, more accurate analyses will need to account more carefully for the unique spatial details of particular real-life urban areas. In addition, this study employed a number of additional assumptions, including (1) that car-to-infrastructure and car-to-car systems have the same impact on accident prevention; (2) that both systems are capable of preventing 100 per cent of all collisions; (3) that car-to-car systems do not require any off-board devices; and (4) that both system on-board units are available at the same price to drivers. In view of these assumptions, this study remains at the conceptual level, and further studies are needed to better reflect the real world.

Perhaps the most important question for future research is how best to combine government policies – given limited financial resources – to optimize net social benefit. In this chapter, we considered the optimal market penetration rates for safety-related ITS on-board units under a wide range of assumptions, but did not take into account the cost of infrastructure installations or the costs of financing and administration. A key objective of our future research is to consider the best possible combination of policy tools such as subsidies, the roadside device installation coverage areas, usage of dual-mode on-board units, and standard equipment provisioning on the part of car-makers.

Notes

1. Institute for Traffic Accident Research and Data Analysis, *Transportation Statistics* (fiscal year 2007 edition).
2. Automobiles travel on the left side of the road in Japan, not on the right side as in the US and continental Europe. However, throughout this chapter we have chosen the sense of left and right turns to place our discussions in a context most familiar to US and European readers. Thus, the type of collision identified as category (3) in the MIC classification is typical of *left* turns in the US and Europe, but of *right* turns in Japan.
3. A *right-turn undercut* collision occurs when a vehicle turning right at an intersection is passed on the right by a fast-moving bicycle or motorcycle. (Again,

as per the previous note, this type of incident typically occurs during *left* turns in Japan.)

4. This value is set in consideration of the average usage lifetime of automobiles.

References

Anas, A. and Xu, R. (1999) 'Congestion, Land Use, and Job Dispersion: A General Equilibrium Model', *Journal of Urban Economics* 45(3): 451–73.

Fujita, M. and Ogawa, H. (1982) 'Multiple Equilibria and Structural Transition of Non-monocentric Urban Configurations', *Regional Science and Urban Economics*, 12(2): 161–96.

He, J., Zeng, Z., and Li, Z. (2010) 'Benefit Evaluation Framework of Intelligent Transportation Systems', *Journal of Transportation Systems Engineering and Information Technology*, 10(1): 81–7.

Katz, M. L. and Shapiro, C. (1985) 'Network Externalities, Competition, and Compatibility', *American Economic Review* 75(3): 424–40.

Katz, M. L. and Shapiro, C. (1994) 'Systems Competition and Network Effects', *Journal of Economic Perspectives* 8(2): 93–115.

Kulmala, R. (2010) 'Ex-ante Assessment of the Safety Effects of Intelligent Transport Systems', *Accident Analysis & Prevention* 42(4): 1359–69.

Leviäkangas, P. and Lähesmaa, J. (2002) 'Profitability Evaluation of Intelligent Transport System Investments', *Journal of Transportation Engineering* 128(3): 276–86.

Libenstein, H. (1950) 'Bandwagon, Snob and Veblen Effects in the Theory of Consumer's Demand', *Quarterly Journal of Economics* 64(2): 183–207.

Ministry of Internal Affairs and Communications (MIC) (2009) 'ITS musen shisutemu no koudoka ni kansuru kenkyu kai houkoku sho' (Report of the Research Group on the Evolution of High-level Wireless ITS Technology), available at http://www.soumu.go.jp/menu_news/s-news/14422.html (accessed December 1, 2010) (in Japanese).

Ministry of Land, Infrastructure, Transport and Tourism (2008), 'Koutsuujiko gensyou ben-eki no gentan-i no sansyutsu houhou' (Calculation method for the traffic accident reduction benefit), available at http://www.mlit.go.jp/road/ir/ir-council/hyouka-syuhou/4pdf/s2.pdf (accessed December 1, 2010) (in Japanese).

Rohlfs, J. (1974) 'A Theory of Independent Demand for a Communications Service', *Bell Journal of Economics and Management Science* 58(1): 16–37.

Spyropoulou, I., Penttinen, M., Karlaftis, M., Vaa, T., and Golias, J. (2008) 'ITS Solutions and Accident Risks: Prospective and Limitations', *Transport Reviews* 28(5): 549–72.

Vaa, T., Penttinen, M., and Spyropoulou, I. (2007) 'Intelligent Transport Systems and Effects on Road Traffic Accidents – State of the art', *IET Intelligent Transport Systems* 1(2): 81–8.

8
Transport Problems and Policy Solutions in China

Sun Lin

8.1 Introduction

The years since the dawn of the twenty-first century have been an era of rapid growth in the production and consumption of vehicles – primarily passenger cars – in China. Before this time, demand for passenger cars had been driven largely by government and state-owned enterprises, but the vehicle market underwent great changes after China's entry into the WTO (World Trade Organization) in December 2001. In anticipation of reduced tariffs on imported vehicles, price competition gradually crept into the domestic market; meanwhile, a segment of the consumer population, having accumulated some disposable income during China's long period of sustained economic growth, began to exhibit surging demand for passenger cars. These favorable trends in both supply and demand have resulted in a significantly enlarged market for passenger cars, primarily sedans, since the year 2002. In 2009, sales of new vehicles exceeded 13.79 million, with passenger cars accounting for 10.33 million, or more than 75 per cent of total sales. By the end of 2009, civilian vehicles in China numbered some 62.88 million.[1]

The size of China's new vehicle market has exceeded that of the Japanese market since 2005 and that of the US market since 2009. However, China still exhibits a very low rate of vehicle ownership compared with developed countries: the ownership rate of passenger cars in China was just 4.65 per cent at the end of 2009.[2] On the other hand, according to analyses and estimates by some experts, the ownership rate of passenger cars among Chinese urban families will exceed 20 per cent by 2011, at which point China will truly enter the era of democratization of passenger cars.[3]

With the rapidly growing automobile market and the increased number of vehicles comes steadily increasing demand for automotive fuel, both gasoline and diesel. This increase in automotive fuel consumption has led to increased emissions of vehicular pollutants, which are the primary contributors to air pollution in both urban and rural areas. In short, all the negative externalities of automobile transportation that had previously been experienced by developed countries are now gradually becoming significant in China.

Although none of these problems has been conclusively eradicated even in the most advanced societies, most first-world nations have succeeded in alleviating the core problems of vehicular transportation through some combination of legislation, technological regulation, and tax, fiscal, and subsidy policy. Though China currently has a much lower vehicle population per capita than the developed countries, it is clearly witnessing a rapid increase in vehicle population. It is thus critical that China pay careful attention to the problems mentioned above, learn from the experiences of the developed countries, and formulate and enact policies suitable for Chinese society as soon as possible.

At present, although comprehensive analyses and assessments of Chinese technological policies related to automotive transportation are rare, some questions have been addressed, by both quantitative and qualitative studies. Wu (2005) analyzed energy conservation policies in China and pointed out that the critical goal for the future is to establish comprehensive policies that encompass laws, regulations and taxes. Xi and Chen (2006), Zhang *et al.* (2007), and Wang *et al.* (2007) conducted quantitative empirical studies to analyze the impact on CO_2 emission reductions of policies designed to encourage alternate means of transportation. Kobos *et al.* (2003) considered questions of energy demand and increased CO_2 emissions in an analysis of the relationship between growing income among Chinese citizens and increasing numbers of passenger cars. Finally, in an application of a computable general equilibrium (CGE) model, including a consumer vehicle-type selection model, Sun *et al.* (2006) and Sun (2007) conducted empirical analyses of policies, including 'Limits on Fuel Consumption for Passenger Cars,' 'Consumption Taxes on Passenger Cars,' 'Fuel Taxes,' and 'Subsidies for Eco-friendly Vehicles' in China.

Given the isolated objective of solving the problems produced by automotive transportation, the most radical policies clearly ensure the most effective solutions. In practice, of course, the problems are not this simple, because the full social costs and benefits of implementing each policy must be carefully considered in assessing the impact of that policy. In Japan, for example, Kii (2007) has evaluated the impact of regulatory policies on 2015 vehicle fuel consumption from the perspective of maximal consumer utility. Muto *et al.* (2006) applied a CGE model to conduct an empirical evaluation of Japan's 'Passenger Car Top-Runner Standards' and 'Green Tax' policies, basing the evaluation on an assessment of benefits to society. Clearly, it is no less important for China to evaluate all relevant policies from the perspective of social costs and benefits. However, we wish here to emphasize that, in the particular matter of assessing the impact of automotive transportation policies in China, there are two unique issues that must be taken into consideration.

The first is the matter of the cultivation and development of the Chinese automobile industry. It is only in recent years that China's automobile industry has begun to develop, but the industry has already taken on several increasingly important roles in the Chinese economy: stimulating associated industries, contributing to tax revenue and job opportunities, and improving the technological level of Chinese industry as a whole. Given these circumstances, policies unfavorable to the development and strengthening of the Chinese automobile industry must clearly be neither formulated nor implemented.

The second is the matter of the cultivation and development of China's 'self-owned brands.' Calls for the cultivation and development of 'self-owned brand' vehicles are on the rise in the Chinese automobile market, which is known as the most fiercely competitive market in the world. The problem relates not only to the nation's pride and self-esteem, but also to its pursuit of pre-eminent roles for the marketing and technological development of domestic vehicles, as well as the widespread anticipation of the development of important future export products by the government and the industrial community.

Although we do not focus on a quantitative empirical evaluation of policy effects, in this article we take full account of the peculiarities of China's automobile transportation problems. Given this proviso, this chapter attempts to sort out the aforementioned problems of automotive transportation in China, conducts an introductory qualitative evaluation of the relevant policies implemented in China so far, and discusses future trends in technological policies based on an understanding of the experiences of developed countries in coping with the same problems. In this article, we begin by analyzing the increase in automotive fuel consumption and the corresponding policy measures taken by the government. Next, we consider the problem of vehicular pollutant emissions and examine the relevant government policies pertaining to this question. Third, we discuss traffic congestion problems and applications of ITS (Intelligent Transport System) technologies. Finally, we summarize the essential characteristics of the policies and policy environment of developed countries and assess the outlook for the corresponding policies in China. We regret that restrictions on the length of this article preclude us from discussing issues related to automotive safety.

8.2 The problem of rising automotive fuel consumption – and the government's policy responses

8.2.1 The rise of automotive fuel consumption in China

China has become the second largest oil-consuming nation in the world, trailing only the US. Although China is an oil producer, in 1993 it crossed over from a net exporter to a net importer of oil, and the quantity of oil it imports and produces has increased every year since. In 2009, China's net imports reached 198.62 million tons and amounted to over 50 per cent of total oil consumed; the nation has become the third largest oil-importing country in the world, behind only the US and Japan. The increased volume of imported oil is attributed primarily to the growth in vehicle fuel consumption, which accounts for about two-thirds of the fuel consumption increase in recent years. In 2007, vehicle fuel consumption was 118.06 million tons, with 97 per cent of all gasoline and 51.3 per cent of all diesel consumed by vehicles (Zhang, 2009).

And yet the volume of vehicle fuel consumption has only just begun to grow. The Development Research Center of the State Council predicts that vehicle sales in China will reach 21.84 million (including 13.77 million passenger cars) by 2020, and, with a vehicle population of 130–150 million, China will be the world's largest consumer of vehicles. China's oil consumption will grow to between 450 and 610 million tons in 2020, of which the fraction consumed by vehicles will rise from 25 per cent in 2005 to about 60 per cent.[4] In addition, various experts estimate that average per vehicle fuel consumption in China is currently still higher than in the developed countries. If we extrapolate based on the current rate of increase in vehicle demand, the vehicle fuel demand will grow at an annual rate of 6 per cent, and vehicle fuel consumption in 2030 will be five times that in 2000 (Jin, 2005). On the other hand, the present low average level of vehicle fuel economy in China indicates that there is significant room for improvement in this area.

The low average level of vehicle fuel economy in China is generally attributed to the poor state of technological advancement that the

Chinese automobile industry took as its starting point, as well as to the Chinese government's protectionist policies for the automobile industry. In the past few decades, the Chinese government has not introduced either strict limits on fuel consumption or tax or subsidy policies to encourage the development of energy-saving vehicle technologies, as has been common in developed countries.

The steadily growing vehicle market and rising vehicle fuel consumption have led to much present anxiety over China's energy security, especially regarding the security of oil, for which China is now increasingly dependent on imports. The questions of how to improve overall average vehicle fuel economy and how to control the increase in vehicle fuel consumption have thus long since taken root as the most crucial developmental issues for China's automobile industry. In recent years, the Chinese government has also come to appreciate the seriousness of these problems, and is now actively attempting to learn from the experiences of developed countries; the government has begun to formulate and implement policies and measures to promote the development and market penetration of energy conservation technologies for the automobile industry.

8.2.2 Policies relevant to energy conservation

Vehicle fuel consumption is determined by several factors, including vehicle population, average fuel consumption, average mileage, and driving speed. Reducing consumption thus requires multi-faceted initiatives. In order to promote the popularity of vehicular energy conservation technologies and improve average fuel economy, policy measures should be designed to stimulate industrial players to develop, commercialize, and mass-produce energy conservation technologies. At the same time, in order to increase the overall level of fuel economy among all active vehicles, additional policy measures must be put in place to accelerate the elimination of old types of vehicles and to reduce their fuel consumption. Fuel taxes, which increase the cost of driving, are also an effective means of decreasing the total distance traveled by existing vehicles. The Chinese government has, in recent years, implemented several policy regulations and technological standards aimed at vehicular energy conservation technologies, which we now briefly review.

The 'Law of the People's Republic of China on Conserving Energy,' enacted in November 1997 and revised in 1998, encouraged the development and dissemination of energy-conserving technologies, but did not directly mention vehicular energy conservation technologies. A revised version, promulgated in April 2008, added a supplement relevant to automotive transportation.

The 'Tenth Five-Year Plan for the Automobile Industry,' released in 2001, mandated that, by the end of 2005, vehicle fuel consumption per 100 kilometers must decrease by 10 per cent versus 2000 levels, with a 5–10 per cent decrease for light-duty vehicles and a 10–15 per cent decrease for medium- and heavy-duty vehicles.[5] The plan also discussed (1) possible research into reforming the vehicle consumption tax, the vehicle purchase tax, and the implementation of the fuel tax,[6] (2) the development of electric and gas–electric hybrid vehicles, and (3) the popularity of alternative-fuel resource vehicles.

The 'Policy on the Development of the Automotive Industry,' enacted in May 2004, proposed, for the first time, a numerical target for improved vehicle fuel economy: by 2010, average vehicle fuel consumption was to decrease by at least 15 per cent relative to the 2003 level, although the policy set no specific limits.

In 2004, specific limits on fuel consumption for passenger cars were finally imposed, and two revisions to the consumption tax structure – a tax increase on large-displacement vehicles and a tax reduction on small-displacement vehicles – were implemented. In addition, regulated limits on fuel consumption by light-duty commercial vehicles were put into practice in February 2008.

a) Standards on 'Limits on Fuel Consumption for Passenger Cars'.

Standardized 'Limits on Fuel Consumption for Passenger Cars' went into effect in October 2004 (see Table 8.1). New vehicles were required to comply with the Stage One fuel consumption limits as of July 1, 2005, and with the Stage Two limits as of January 2008. Previously approved vehicles were required to comply with the Stage One limits by July 1, 2006, and with the Stage Two limits by January 1, 2009. Stage Three will be released by the end of 2010.

The goals of the standard are to stimulate the automobile industry to improve vehicle engine performance, to enhance the overall fuel economy levels of all vehicles, to encourage consumers to purchase passenger cars that consume less fuel, and to control the trend of continuous increases in vehicle fuel consumption. The standard levels will be gradually tightened in the future. In conjunction with these regulations, it has become official practice in China to publicize the fuel economy levels of both newly approved and previously approved vehicles.

Table 8.1 Standards imposed by the 'Limits on Fuel Consumption for Passenger Cars'

Vehicle Weight (CM) (kg)	Stage 1	Stage 2
CM ≤ 750	7.2	6.2
750 < CM ≤ 865	7.2	6.5
865 < CM ≤ 980	7.7	7.0
980 < CM ≤ 1,090	8.3	7.5
1,090 < CM ≤ 1,205	8.9	8.1
1,205 < CM ≤ 1,320	9.5	8.6
1,320 < CM ≤ 1,430	10.1	9.2
1,430 < CM ≤ 1,540	10.7	9.7
1,540 < CM ≤ 1,660	11.3	10.2
1,660 < CM ≤ 1,770	11.9	10.7
1,770 < CM ≤ 1,880	12.4	11.1
1,880 < CM ≤ 2,000	12.8	11.5
2,000 < CM ≤ 2,110	13.2	11.9
2,110 < CM ≤ 2,280	13.7	12.3
2,280 < CM ≤ 2,510	14.6	13.1
2,510 < CM	15.5	13.9

Notes: Unit: L/100km.
Source: http://www.fueleconomy.cn/doc_pdf/chych-bzh.pdf" (accessed May 11,2011).

b) Revisions to consumption tax rates on passenger cars and fuel consumption taxes.

Consumption tax rates on passenger cars in China are established in conjunction with vehicle air displacement standards. Consumption tax rates were adjusted for the first time in April 2006. As shown in Table 8.2, the tax rates for vehicles with engine displacement of 1.5 l decreased, the rates for vehicles with displacements of 1.5–2.0 l remained unchanged, and the rates for large-displacement vehicles, and particularly for vehicles with displacements of more than 3.0 l, increased sharply, indicating a clear policy intention to favor small-displacement vehicles while penalizing those with large displacement. A second revision, in September 2008, again sharply increased the tax rate for vehicles with displacement greater than 3.0 l, while further reducing taxes on vehicles with displacement of 1.0 l or less.

In December 2005, before the April 2006 revision to the vehicle consumption tax rates, the State Council released a document entitled 'Perspectives on Encouraging the Development of Small, Energy-Saving,

Table 8.2 Revisions to consumption tax rates on passenger cars in China (%)

Air displacement (Unit: L)	Tax rate before April 2006 revision	Tax rate after April 2006 revision	Tax rate after September 2008 revision
L ≤1.0	3.0	3.0	1.0
1.0< L ≤1.5	5.0	3.0	3.0
1.5< L ≤2.0	5.0	5.0	5.0
2.0< L ≤2.5	8.0	9.0	9.0
2.5< L ≤3.0	8.0	12.0	12.0
3.0< L ≤4.0	8.0	15.0	25.0
4.0< L	8.0	20.0	40.0

Source: Prepared by author based on data from the website of the Ministry of Finance of the People's Republic of China (http://www.mof.gov.cn/).

Eco-Friendly Cars,' which requested local governments across China to repeal regulations prohibiting small-displacement vehicles – particularly micro-passenger cars with displacements less than 1.0 L – on arterial roads. The document requested that local governments repeal such regulations before March 2006, with an eye toward accelerating the evolution of passenger cars toward small-displacement vehicles, as had already occurred in the developed countries.

However, after the April 2006 tax revisions, no such evolution was observed. Indeed, on the contrary, the market share commanded by small-displacement vehicles appeared to be steadily shrinking in 2007.[7] Due to the advantages of low price and low fuel consumption, it was generally believed that micro-passenger cars would first be popularized in China. However, in actual fact, many of the micro-passenger cars produced in China have high operating fuel consumption and fail to meet minimal safety performance standards. Moreover, social considerations contribute to a general preference in China for vehicles with larger air displacement. All of these factors have resulted in a steadily decreasing share for micro-passenger cars in the Chinese automobile market. In view of the Chinese consumer's preference for large-displacement vehicles, we anticipate that micro-passenger cars will vanish from the Chinese market unless stronger policy measures are introduced to stimulate technological improvements in the fuel economy, pollutant emissions, and safety performance of these vehicles.[8]

Before 2009, there were few policies designed to accelerate the abandonment of old, inefficient vehicle types. Policies to subsidize this acceleration, to encourage the elimination of old models, and to subsidize the vehicles that meet energy-saving standards were first introduced in 2009. Around the same time, fuel tax policies – which had never been imposed despite having been studied for many years and clearly identified as an effective means of increasing the cost of large-displacement, high-fuel-consumption vehicles – were finally implemented. On December 19, 2008, the Chinese government issued a 'Notice of Implementing the Reform on the Price and Tax of Refined Oil,' which imposed a system of refined-oil taxes and price reforms effective as of January 1, 2009.

The purpose of these reforms was to establish market price mechanisms for refined oil, to standardize the tax and fee systems in the transport sector, to promote energy conservation, to ensure that the burden of constructing and maintaining roads would be fairly distributed among different type of users based on their usage of the roads, and to accumulate funds for transport infrastructure maintenance and construction.

The reform did not establish a new type of fuel tax, as had been expected. Instead, it increased the tax per unit of the existing Fuel Consumption Tax and eliminated some associated road tolls. The primary components of the reform included: 1) elimination of six waterway- and road-related charges, including road tolls; 2) a phasing out of tolls on secondary roads, which repaid local government loans; 3) an increase in the fuel consumption tax per unit, with the gasoline tax increased by RMB 0.2 to RMB 1.0 per liter, the diesel tax increased by RMB 0.1 to RMB 0.8 per liter, and taxes on other oil products increased accordingly.

The Fuel Consumption Tax is levied in the production phase, based on the quantity of fuel consumed with a fixed charge per unit, and is collected together with the price of the fuel. Revenue from this tax is general revenue, belonging to central government. However, additional revenues resulting from the increased fuel consumption tax are not treated as general revenue, but are instead transferred to local governments to set up transportation funds. These funds are then used (1) to replace the income that had previously been derived from the six charges (such as road tolls) eliminated by the 2008 reforms, (2) to subsidize government loan repayments for secondary roads on which tolls had previously been collected, and (3) to subsidize the grain producers and the urban public transport sector, including taxis.

Initially, the reform of the Fuel Consumption Tax was well received by the Chinese public. However, people gradually came to realize little or none of the benefit that had been expected from the reform. Some local governments did not abolish secondary road tolls and city road tolls as early as had been anticipated. Also, because of the monopoly in China's refined oil market, refined oil prices rose rather than fell. Currently, the prices of gasoline and diesel oil are far in excess of their levels in 2008, when international crude oil prices peaked. In addition, the number of public vehicles that depend on public finance did not decrease at all. Some of these factors increased the burden on private owners, while others offset the intended objectives of the increased fuel consumption tax.

8.2.3 Evaluation of policies enacted

Although the average fuel consumption of vehicles in the Chinese market has improved rapidly in recent years, it is still high compared with that in developed countries. It may thus be more important for China to accelerate technology-based improvements in average vehicle fuel consumption than to control vehicle population and mileage. For this reason, China has finally begun to implement policies designed to improve individual vehicle fuel consumption. The 'Limits on Fuel Consumption for Passenger Cars,' enacted on 1 July, 2005, was the first regulation of vehicle fuel consumption; as the first regulation of vehicle fuel consumption in China, it has important policy implications. An additional set of 'Limits on Fuel Consumption for Light-Duty Commercial Vehicles' was also adopted on July 1, 2007. Fuel consumption regulations for gas–electric hybrid alternative-fuel vehicles will also be gradually formulated and imposed.

However, China is still in the process of researching, and learning from, the experiences and achievements of the developed counties in instituting tax preference or allowance policies to promote the popularity of energy conservation technologies. Indeed, the present taxation system contains strong policy initiatives to increase taxes on vehicles with high fuel consumption, but only weak provisions for decreasing taxes on, or providing allowances for, vehicles with small fuel consumption. The existing system, in other words, places strong emphasis on penalties while neglecting rewards. Our earlier discussion of the adjustments made to the vehicle consumption tax structure illustrated that large-displacement vehicles were subject to a significant range of tax increases, while privileges for small-displacement vehicles were negligible.

Energy conservation technologies are significant not only for reducing fuel consumption but also for reducing greenhouse gas emissions. China surpassed the US as the largest emitter of greenhouse gases in 2007. For this reason, although China is not subject to the restrictions of the Kyoto Protocol, the nation nonetheless faces mounting international pressure. In response, the Chinese government in June 2007 announced a 'National Climate Change Program' aimed at reducing greenhouse gas emissions. The program lays out basic principles to remain in effect through 2010, states the key concepts in greenhouse gas mitigation, and imposes overall reduction targets and policy measures. This national initiative sets an overall target of 20 per cent reduction in energy consumption per GDP unit over 2005 levels by 2010. The program also designates transportation as a key strategic field. Policies on taxes and allowances in vehicle-related fields will be designed to reward the development, production, and consumption of energy conservation technologies, eco-friendly small-displacement vehicles, gas–electric hybrid vehicles, and electric vehicles. At the same time, the program will seek to accelerate the abandonment of old energy-intensive vehicles, and standard limits on vehicle fuel consumption will be gradually tightened.

In the foreseeable near future, the issues of energy security and global warming policies will stimulate China to design policies to popularize small-displacement, energy-conserving vehicles. In the future, policies to promote industrial commitment to the development of energy conservation vehicle technologies, and to promote consumer purchases of energy-conserving vehicles, must be designed and implemented.

In comparison with other nations, China was a latecomer to the environmental protection table. Fuel efficiency regulations were not introduced in China until 2004, and these only set *ceilings* on fuel efficiency, as described in Chapter 2; in contrast to developed nations such as Japan and the US, China does not yet regulate the *average* fuel efficiency of vehicles sold.[9] In view of the rapid motorization it is currently experiencing, China must review the regulatory approaches taken in nations such as Japan and the US, and must initiate studies similar to those conducted in Chapters 2 and 3.

8.3 The problem of vehicular pollutant emissions – and the government's policy responses

8.3.1 Air pollution problems in China's urban areas

In China's urban areas, vehicular pollutant emissions have been surging for years, resulting in serious air pollution problems. According to the 'First National Pollution Census Bulletin,'[10] released on February 6, 2010, vehicular emissions in 2007 of particulate matter (PM), hydrocarbons (HC), carbon monoxide (CO), and nitrogen oxides (NO_x) in China reached 0.59, 4.78, 39.47, and 5.49 million tons respectively, and NO_x

emitted from motor vehicles accounted for 30 per cent of total nitrogen oxide emissions in China. Problems of photochemical pollution also occur frequently in some cities. In addition, small particulate matter (SPM), of which vehicular emissions account for 20–30 per cent, has become an important pollutant in many cities. Some experts predict that Chinese vehicular emissions of CO and NO_x by 2010 will be double those in 2004, and the fraction of overall pollution due to vehicles in medium-sized and large cities will reach 79 per cent (Li, 2005).

Let us consider the case of Shanghai as an example. In the last decade or so, the dominant cause of air pollution in this city has shifted from coal burning to the burning of mixed oil and coal. With an increasing population of vehicles, the average concentration of NO_x around Shanghai has been on the rise since 1990, with concentrations in 2004 already 1.75 times those in 1990. The downtown area has always seen a higher average concentration of NO_x than the suburbs. In recent years, however, reinforced controls on vehicular pollutant emissions in downtown areas have contained the increase in the NO_x concentration, while a rising trend remains obvious in the suburbs.

Most cities in China are still in the early stages of the spread of family passenger cars. Based on present popularity rates, vehicle emissions will continue to aggravate air pollution unless strict control measures are taken. Since the year 2000, the Chinese government has begun to realize the severity of this problem, just as it has realized the severity of the problem of steadily rising vehicle fuel consumption. In order to contain the inexorable increases in vehicular pollutant emissions, the government has accelerated its efforts to design and implement policy measures to constrain emissions, promote the development of clean vehicles, and alleviate urban air pollution.

8.3.2 Air pollution policies

In the medium and long term, the most effective means of reducing vehicular pollutant emissions is to promote the development and popularity of clean vehicles, such as gas–electric hybrid vehicles, electric vehicles, fuel-cell vehicles, and biofuel vehicles. However, the most effective measures in the *short* term are to improve the performance of existing fuel engines, to improve the efficacy of pollutant purification devices, and to reduce the engine displacement of mainstream vehicles in the market. In addition, it is also important that policies stimulate the fuel production industry to improve the quality of existing vehicle fuels and to reduce harmful ingredients in fuels. Besides the technological countermeasures mentioned above, other important methods for promoting the popularity of new emission technologies include (a) tax policies that favor technology development enterprises and (b) measures to encourage consumers to replace old, high-emission vehicles with new, small-displacement vehicles. Effective measures for reducing emissions also include improvements in transportation management, restrictions on driving in rush hours and in congested areas, and traffic alleviation.

In recent years, the Chinese central government and some local governments have established certain policies to control increases in vehicular pollutant emissions, including reward policies to encourage technology development and industrialization for clean vehicles, tax policies to favor small-displacement vehicle manufacturers, and regulatory policies to encourage market inhibition of high-emission vehicles. In addition, in big cities, some local governments have taken measures to prevent high-emission vehicles that fail to meet present emission standards from driving on certain roads or at certain times. For example, in 2002, tough administrative measures were taken by the Beijing municipal government to forbid sales, and enforce compulsory elimination, of minibuses with high fuel consumption and high emissions. In Shanghai, measures were taken to forbid vehicles which fail to meet present emission standards from driving on roads in the inner-loop regions of the city.

Policies to constrain vehicular pollutant emissions in China were put into effect before policies relating to energy conservation. By 2000, the policies were in place to encourage the spread of clean-vehicle technology, including legal policies, technology policies, and tax reductions or exemptions for small-displacement, low-emission vehicles. Since the year 2001, specific policy objectives, including the introduction of European standards, implementation of a stricter vehicle inspection system, improvements in vehicle fuel quality, and the construction of a sustainable transportation system, have been articulated. Subsequently, limits on pollutant emissions for passenger cars, motorcycles, farm transport vehicles, and trucks were gradually formulated, announced, and implemented by the government.

(a) Policies to limit vehicular pollutant emissions.

Specific limits on vehicular pollutant emissions date back to the 1980s, but standards on emission limits (the European emission standard) for specific vehicle types were not introduced until 1999. The 'National Standard,' which targeted light-duty vehicles, was formulated and introduced in 2001, and had almost the same regulatory objectives as the 'Technology Policy for the Prevention and Cure of Motor Vehicle Exhaust Pollution,' introduced in May 1999 with reference to the European emission standard. The specific schedule was that Stage One emission standards should take effect in 2000, Stage Two emission standards would take effect in 2004, Stage Three emission standards would take effect in 2007, and Stage Four standards would take effect in 2010 for newly approved vehicles. Stage Two through Four emission standards would go into effect one year later for previously approved vehicles than for newly approved vehicles. However, in Shanghai, Beijing, and Guangzhou, advanced-stage emission standards would go into effect earlier than elsewhere.

The 'Tenth Five-Year Plan for the Automobile Industry,' announced in March 2001, indicated that, by 2005, newly approved light duty vehicles, large and medium buses, and medium- and heavy-duty trucks should strive to meet the Europe II emission standards, while a portion of top-grade passenger cars and large and medium buses should strive to meet the Europe III emission standards. Newly approved farm transport vehicles should meet the Europe I emission standards.[11] By around 2010, all newly approved vehicles must meet international emission standards.

The document 'Limits and Measurement Methods for Emissions from Light-duty Vehicles' contains provisions, measurement methods and content identical to those specified in the 'Technology Policy for the Prevention and Cure of Motor Vehicle Exhaust Pollution,' but augments the regulation standards for vehicles running on gaseous fuels. Regulations for Stages One and Two were introduced in April 2001, while those for Stages Three and Four were introduced in March 2005. Details of the content of, and the schedule for, the Europe I through IV standards are shown in Table 8.3.

The 'Technology Policy for the Prevention and Cure of Diesel Vehicle Exhaust Pollution,' released in January 2003, limits emissions of CO, NO_x, THC, and PM from diesel vehicles, stating that the most crucial goal is to control emissions of NO_x and PM. Newly approved vehicles should meet Europe II emission standards by 2004, and should meet Europe III emission standards by 2008. Programs limiting emissions from other vehicles, such as three-wheel vehicles, low-speed farm transport vehicles with diesel engines, and motorcycles, have also been put into effect since 2003.

Before the imposition of the Europe II emission standard in 2004, the Chinese government, in order to improve levels of technology regarding vehicle emissions, introduced tax policies to stimulate the vehicle

Table 8.3 Progress on limiting vehicular pollutant emissions in China

Standard	Date of implementation in China	Date of implementation in Europe	Time lag (years)
Before EURO	1993	1973	20
EURO I	2001	1992	9
EURO II	2004	1996	8
EURO III	2008	2000	8
EURO IV	2010	2004	6

Note: For an international comparison, see Figure 1 in Seko (2007).

industry, reducing consumption tax by 30 per cent on vehicles with fewer than 22 seats, in accordance with the Europe II emission standard. From 2001 to the end of 2003, a total of 42 manufacturers and 1,043 types of vehicles were rewarded with tax reductions, of which the total sum amounted to RMB 8 billion.

Generally speaking, when it comes to limiting vehicular pollutant emissions, China lags behind developed countries, but takes faster steps. It took Europe 27 years to reach the Europe III emission standard, while China is predicted to reach this level in 17 to 19 years. Estimates suggest that each transition to a more stringent standard on emission limits reduces harmful gas emissions by 30 per cent per vehicle. Experts predict that, if the Europe III and IV emission standards are implemented as planned, annual emissions of NO_x, HC, and CO by vehicles on ordinary roads will be reduced by 1.9, 2.2, and 16 million tons, respectively. In addition, the nine-year gap in limiting vehicular emissions that existed between China and the European countries in 2001 is predicted to shrink to six years in 2010 (Li, 2005).

b) Policies to stimulate the development of new-energy vehicles.

In its 'Policy on the Development of the Automotive Industry', introduced in 2004, the Chinese government proposed (a) the development of energy-conserving and eco-friendly small-displacement vehicles, (b) research and development for, and industrialization of, alternative-fuel vehicles such as electric vehicles and fuel-cell vehicles, (c) steps to promote the popularity of gas–electric hybrid vehicles, and (d) steps to encourage the vehicle industry to research, develop, and mass-produce vehicles using new energy sources such as methanol, ethanol, natural gas, blended fuel, and hydrogen fuel.

Before the implementation of the 'Tenth Five-Year Plan for the Automobile Industry,' the term 'clean vehicles' in China mainly referred to vehicles using gaseous fuels such as compressed natural gas (CNG) and liquefied petroleum gas (LPG). In 2005, gaseous-fuel vehicles numbered around 220,000 and accounted for around 50 per cent of all buses and taxis. After the introduction of the 'Tenth Five-Year Plan for the Automobile Industry,' the Chinese government added research projects on gas–electric hybrid vehicles, electric vehicles, fuel-cell vehicles, and biofuel vehicles to the key projects supported by the 'National High-tech R&D Program of China – 863 Program,'[12] and pumped RMB 880 million into research departments to support research and development for new technologies during the term of the 'Tenth Five-Year Plan.'

The 'Eleventh Five-Year Plan for the Automobile Industry' designated research and development, and industrialization, for energy conservation and clean-vehicle technologies as critical national projects, and proposed guidelines for stronger support of gas–electric hybrid vehicles, electric vehicles, fuel-cell vehicles, and biofuel vehicles. The overall objective was to realize the commercialization of fuel-cell vehicles and the industrialization of gas–electric hybrid vehicles during the period of the 'Eleventh Five-Year Plan,' that is, by the end of 2010. Standards for the safety and power performance of electric vehicles were gradually formulated and introduced. Meanwhile, technological standards for the safety, power performance, and fuel consumption of some light-duty and heavy-duty gas–electric hybrid vehicles, as well as techniques for measuring pollutant emissions from those vehicles, were formulated and introduced between May and October 2005. Standards for gas–electric hybrid vehicles, electric vehicles, fuel-cell vehicles, and biofuel vehicles are currently under rapid development.

Since 2007, policies relating to the new-energy vehicle industry have gradually evolved to include greater support of industrialization programs, beyond the original goals of pure research; such policies include subsidies, demonstration projects, a tightening of market entry regulations, and other measures.

In particular, during the 'Tenth Five-Year Plan,' the growth of non-state-owned enterprises and the increasing number of firms developing new-energy vehicles led government policies eventually to assume a hybrid form, in which a market-based regulatory system was combined with incentive schemes. For instance, a document entitled 'Rules Regarding Market-Entry Regulations on the Production of New Energy Automobiles' was issued in November 2007 to clarify market entry regulations. This document defines new-energy vehicles and imposes certain conditions and requirements that new-energy vehicle manufacturers must meet before entering the market. With respect to incentive policies, a set of 'Interim Measures to Provide Financial Assistance for the Promotion of Energy-Saving and New-Energy Vehicle Demonstrations' was introduced. These measures define clear conditions and subsidies for a variety of new-energy vehicles, with the objective of encouraging manufacturers to improve the technological level of their new-energy vehicles. A 'Notice on Pilot Programs, Promotions, and Demonstrations of Energy-Saving and New-Energy Vehicles' was issued to strengthen the government's mandatory procurement of energy-saving and environmentally friendly vehicles and their promotion within the public transportation sector. A 'Directory of Recommended Types of Energy-Saving and New-Energy Vehicle Demonstration Projects for Promotion Applications' was issued to stimulate the creation of consumer demonstrations.

In November 2008, China's Ministry of Science and Technology kicked off a promotion campaign entitled '1,000 Vehicles in Ten Cities.' This program will introduce 1,000 new-energy vehicles, for demonstration purposes, in each of ten cities per year, over a period of three to four years; by the end of the program, at least 60,000 energy-saving and new-energy vehicles will be introduced into various fleets – including those used for public transportation, taxis, civilian vehicles, municipal vehicles, and postal delivery – within the large and medium-sized cities targeted.

In January 2009, a document entitled 'Notice on New-Energy Vehicle Demonstration and Promotion' was issued; this marked the launch of subsidies for the purchase of new-energy vehicles in the public transportation sector, with the objective being to promote the formation of a new-energy auto market. To date, 13 cities have been targeted for promotion demonstrations and other experimental programs. The central government has agreed to provide a one-time fixed subsidy to the pilot cities for the purchase of hybrid vehicles, pure electric vehicles, and fuel-cell or other energy-saving new-energy vehicles. In March 2009, a document entitled 'Auto Industry Restructuring and Revitalization Planning' was introduced; this document proposed an increase in production capacity to 500,000 new-energy vehicles, accounting for some 5 per cent of total vehicle sales by the end of 2011. In May 2009, the State Council granted discounts, in the form of loan arrangements, of RMB 20 billion to the auto industry to stimulate technological innovation in the development of new vehicles.

8.3.3 Evaluation of policies enacted

Since 2000, the Chinese government has introduced many policy measures aimed at containing growth trends in air pollution due to vehicular pollutant emissions. One important component of these policies is tighter limits on pollutant emissions from vehicles already in operation. Another is the medium- and long-term perspective emphasized by the government's support for the development and popularization of future clean vehicles, including gas–electric hybrid vehicles, electric vehicles, fuel-cell vehicles, and biofuel vehicles.

Limits on emissions from vehicles operating on present-day fuel sources have also continued to progress, and the gap of standard limits between China and the developed countries has gradually narrowed. The improved technological capacity of the Chinese vehicle industry to combat pollutant emissions, and the actual reduction in pollutant emissions, testify to the growing impact of government policies. In the future, more rigorous measures must be taken to promote vehicle inspections and the abandonment of obsolete vehicles. In addition, since vehicle emissions closely track traffic congestion, efforts to formulate policy measures to alleviate the problems of traffic congestion in urban areas must be intensified.

Due to policy support from central government, local governments are taking active measures to support the research and development, commercialization, and industrialization of new-energy vehicles, buses, and motorcycles, and have made great progress. Over 500 gas–electric hybrid, electric, and fuel-cell vehicles were put into operation during the 2008 Beijing Olympic Games, and during the 2010 Shanghai World Exposition more alternative-energy bus lines were put into operation.

But specific policies remain rare. All major domestic vehicle enterprises have established joint ventures with multinational corporations, which thus far have essentially held all power in supervising the development of new types of passenger cars. The research and development capabilities of China's domestic vehicle industry thus remain weak. It is for precisely this reason that Chinese government support for new-energy vehicle development is targeted primarily at research and development by universities and national research laboratories. Of course, policies have changed significantly since 2009, with perhaps the most important change being the growth of subsidies to promote new-energy vehicles. Since July 2010 these subsidies, in tandem with standards regulating energy efficiency, have expanded from the public sector to encompass private-sector vehicles as well. We predict that programs to subsidize new-energy vehicles will be further expanded and improved in future.

8.4 Traffic congestion problems and ITS

8.4.1 Traffic congestion problems in China

With continuous, stable economic growth and large-scale development of cities ongoing, both the sizes and populations of cities have risen steadily in China. This rapid urbanization, and the sharply increasing urban population, pose significant challenges for road traffic in metropolitan areas; these challenges have been particularly acute in the past five years, as the population of vehicles, led by family passenger cars, has skyrocketed. Road construction has lagged behind increasing vehicle populations in all cities, resulting in continual declines in travel speed in cities. Traffic congestion has become the norm in cities, and particularly in big cities. Traffic flows at the intersections of arterial roads in Beijing have long since exceeded their limits, frequently leading to travel speeds of less than 5 km/h during rush hours.

At present, urban bus systems carry less than 10 per cent of all passenger transportation. Traffic congestion degrades the quality of bus transit, thus motivating urban residents in non-downtown areas to purchase and use private vehicles. This in turn further aggravates traffic congestion, and a vicious circle ensues. In addition, the lack of parking places for public and commercial facilities in downtown city areas has been further strained by the steady increase in private vehicles, leading to rampant illegal parking, which again further aggravates traffic congestion.

Urban traffic congestion leads not only to deteriorating quality of public transportation services, but also to increased fuel consumption, pollutant emissions, and traffic accidents. For these reasons, China's central government and local authorities have established a set of policies to promote intensive development of rail transportation systems, such as subways and overhead railways, as well as public transportation systems such as bus rapid transit (BRT). In addition, a variety of congestion prevention measures have been considered and implemented in some of China's big cities, including collection of parking fees, driving prohibitions based on the parity (even or odd) of the vehicle's license plate number in Beijing, restrictions on private vehicle ownership by auctions for motor vehicle license plates in Shanghai, and the introduction and development of ITS technologies.

8.4.2 ITS in China

In September 2005, the State Council of China released a 'Notice of Recommendations on Prioritizing the Development of Urban Public Transport,' which proposed five policy measures to address problems in public transportation:[13] financial regulations, networks of priority lanes for buses, signaling systems, a comprehensive transportation information platform, and an intensification of efforts to develop ITS standards.

The 'Program Outline for the Road Transportation Industry (2001–2010),' released in January 2001 by the Ministry of Communication, called for measures such as promoting the digitization of transportation information, encouraging the dissemination of transportation information technologies, including Global Positioning System (GPS) and Geographic Information System (GIS), and encouraging research and development for ITS technologies. Subsequently, the 'National Program Outline for Medium- and Long-Term Scientific and Technical Development (2006–2020),' released in February 2006 by the State Council, deemed ITS technology a developmental priority. This document also required improvements in domestic Chinese capacity for development of ITS-related technologies and established basic policies for the development of ITS technologies.

The 'Eleventh Five-Year Plan for State Science and Technology Development,' which was enacted in November 2006 by the Ministry of Science and Technology based on the core policies of the State Council, included an accelerated development plan for ITS technologies. The main provisions included the establishment of a national pilot project for ITS technologies, the addition of 'modern transportation technology' to the 'National High-Tech R&D Program of China – 863 Program' in 2006, and the launch of a comprehensive ITS research project. Key items mandated that ITS research and development be promoted in the largest cities, including Beijing, Shanghai, and Guangzhou; that a pilot ETC project be implemented in the Yangtze River Delta area and in the Bohai Rim; that technologies and products to improve traffic safety performance be intensively developed; that financial support and tax preferences be provided for research and products within the ITS field, in order to promote the industrialization of ITS; and that a comprehensive standardization of ITS technologies be implemented.

Standards for ITS technologies have been maintained by the Standardization Administration of China since September 2003, and

12 national standards on ITS-related technologies were announced in May 2006 and have been in place since April 2007.

In the field of ITS-related technology applications, even China's largest cities, Beijing, Shanghai, and Guangzhou, still have some catching-up to do when compared with the situation in developed countries. Some experts estimate that the development level of ITS in Shanghai at the end of 2005 was roughly on a par with that of Japan or the US in 1990 – in other words, that ITS development in Shanghai lags behind that in the developed world by some 15 years (Yokota, 2004).

In order to alleviate traffic congestion, the Shanghai municipal government has accelerated large-scale public transit construction, with the primary focus on subways and on construction of ITS-related facilities to increase vehicle travel speeds.

In the field of electronic toll collection, a 'one card solution' system for Shanghai public transportation has been instituted to collect fees for subways, buses, highways, and parking, achieving a level of sophistication of international caliber. However, in general, ETC in China is still in the technological research phase, and is thus somewhat backward when compared with the developed world. As for intelligent public transportation systems, the 'Eleventh Five-Year Plan' calls for the popularization of onboard GPS terminals and systems for displaying the names of bus stations in Shanghai's public transit system before the end of the five-year period. However, wide gaps still exist in many aspects of operational dispatch and scheduling for public transportation systems.

In the field of dynamic mobile navigation systems, Shanghai's government is still at the stage of collecting information. Transportation information-collection and guidance systems, in the form of variable information boards, have been set up on express roads and arterial roads in the central area of the city, but dynamic traffic information announcements aimed at onboard terminals have not yet commenced operation, in contrast to the VICS system in Japan and the European Radio Data System–Traffic Message Channel (RDS–TMC) system. A comprehensive traffic information platform project was approved and initiated in 2006 – some 10 years later than the launch of Japan's VICS center in 1996, but contemporaneous with systems in Beijing and Guangzhou.[14]

The Shanghai municipal government, taking advantage of the opportunities presented by the 2010 Shanghai World Exposition, is further promoting the introduction of traffic information and intelligent decision-making technologies, and has long-term objectives for applications of ITS-related technologies. The long-term objectives include the construction of a platform for urban traffic information sharing and exchange, which could implement effective sharing of various traffic information and strive to become one of the critical information centers in Eastern China and perhaps in the entire country; to enhance the effectiveness of traffic information management until it compares to that in the most advanced cities in the developed world; and to make the intelligent transport industry a center of industrial development and growth in the city.

8.4.3 Evaluation of policies enacted

More powerful promotional policies are needed to stimulate the rapid standardization and dissemination of ITS technologies. However, when it comes to practical implementation, China still lags considerably behind Japan. Of course, China has the advantage of using the newest technology. In Japan, for systems such as the VICS system discussed in Chapter 6 and the Roadside Information-Based Driving Support System discussed in Chapter 7, the use of roadside infrastructure devices has been advanced and practical, both for congestion mitigation and for safety enhancement. However, there is no need for China to adhere to the same path. Inter-vehicle-type communication systems, such as the Probe-Car Information System and the Inter-Vehicle Communication Type Driving Support System, may also be considered as the primary ITS systems to design in to China's future transit infrastructure. The questions of which type of system to adopt, and what policy measures to use to promote it, may be considered under the theoretical framework discussed in Chapters 5 and 7.

8.5 Conclusions

In the following paragraphs we will discuss some directions for technological policies in China, with reference to the experiences and policy histories of developed countries such as Japan.

The experiences of developed countries such as Japan illustrate that technological policies addressing vehicle transportation problems may be broadly divided into two categories: policies on laws, regulations, and technological standards, and policies on taxes, financial rewards, and financial penalties. Such policies have the following general features. First, the levels of regulated technological standards should increase gradually in a predetermined sequence of stages. Second, fiscal subsidies should be implemented at the highest-cost stages of new technology development and dissemination. Third, tax measures should include both rewards and penalties and should be revenue-neutral. Fourth, technological development is carried out by automobile manufacturers, and the impact of the government is felt in technological regulations. Fifth, tax and fiscal policies should generally be directed at consumers rather than directly at industrial players. Finally, technological gaps between automobile manufacturers are relatively narrow, and the government does not need to formulate technological policies in consideration of the particular conditions of individual firms (whether foreign-funded or domestic-funded).

At present, China's policies cannot be directly compared with the corresponding policies of developed countries. But the experience of the developed countries suggests that, above all, China must recognize the

Table 8.4 Major policies aimed at alleviating the externalities of vehicle transportation

Problem addressed	Law and administrative regulation	Technological regulation and standard	Finance and revenue
Vehicle fuel consumption (energy conservation and global warming)	Law of Energy Conservation	Limits on the fuel consumption of vehicles	Adjustments to consumption taxes Support for development of new-energy-technology vehicles Collect fuel taxes
Vehicular pollutant emissions (air pollution)	Law on the Prevention and Control of Atmospheric Pollution Travel prohibitions on unapproved vehicles	Limits on pollutant emissions from vehicles	Support for development of new-energy-technology vehicles
Traffic congestion	License plate restrictions on private vehicles (Shanghai) Parking restrictions (revised parking fees) Travel restrictions (in specified times and areas)	ITS standards and execution of the ITS pilot projects	Support and enforce construction of rail transit systems, BRT systems and ITS

Source: Compiled by author.

importance of forming a comprehensive policy system within which to craft technological policies. In the past, the Chinese government has preferred to establish various technological standards and benchmark regulations (see Table 8.4), but, in the future, comprehensive research on fiscal and tax policies must be conducted as part of a comprehensive scheme that combines technological regulations with tax and fiscal penalties and rewards.

Second, although at the initial stages of the development and dissemination of new technologies, fiscal subsidies and tax incentives can motivate producers and consumers through pricing mechanisms and can generally produce positive effects, the situation in China is somewhat complicated. As mentioned above, since the technological capabilities of 'self-owned brand' vehicle manufacturers are relatively weak, the development of new technologies is essentially overseen by the government and carried out in the form of national research projects at universities, national research laboratories, and major state-owned vehicle enterprises with government financial support. Therefore, if the Chinese government applies fiscal subsidies to new technologies developed by domestic-funded or foreign-funded manufacturers, irrespective of the capital structure of the research institutions or manufacturers, then this could engender some resentment. Particularly today, with ever greater expectations for government cultivation of domestic 'self-owned brands' in the automobile industry, it may be difficult for people to accept fiscal subsidies, and tax incentives and penalties, for new technologies when those new technologies are controlled by the foreign sides of joint ventures or multinational corporations. The policy-making departments must fully consider the technological gap between China's 'self-owned brand' enterprises and multinational corporations, and must enact policies in stages, particularly policies on fiscal subsidies and tax incentives.

A third problem is the non-uniformity of technological regulations among different policy-making departments. The National Development and Reform Commission holds that problems arising in the dissemination of energy-related and environmentally friendly vehicle technologies should be solved by improvements to the structure and technological capabilities of the vehicle industry. This agency thus promotes gradual tightening of technological regulations based on the technological capacity of 'self-owned brand' vehicle manufacturers. On the other hand, in its efforts to improve the environmental situation as quickly as possible, the Ministry of Environmental Protection prefers instead to strengthen existing environmental regulations and hopes to accelerate technology regulations. The Ministry of Science and Technology treats the problems of technological regulations from a medium- and long-term perspective, and pays more attention to research and development for futuristic technologies. Differences in approaches among different agencies are to be expected; however, they affect the selection of which technologies to pursue and how to regulate them.

A fourth problem is the structure of China's vehicle-related tax hierarchy. In many countries, the primary vehicle-related taxes are fuel taxes – that is, taxes collected during the operating phase. In contrast, China has traditionally emphasized taxes on manufacturing and sales; indeed, such taxes comprise one-third of the sales price of passenger cars in China. Although China has recently begun to collect fuel taxes, as discussed previously, the primary emphasis of vehicle-related taxation remains in the manufacturing and purchase phases. Such a tax system is clearly inefficient for the development of the vehicle industry and the alleviation of vehicle transportation problems. It is thus important for China to shift vehicle-related taxation away from the manufacturing and purchase phases and instead toward the operating phase.

Finally, we consider problems with the technology development system. In China, the development of new technologies, and particularly futuristic technologies, is usually carried out by national research institutions and universities with financial support from the government. However, the efficiency of such a technology development system has been questioned. Aside from fundamental research, China should abandon its system of government-directed technology development and should instead establish a system in which companies and research institutions conduct their own development operations. This readjustment should apply in particular to the case of China's transportation-related technology development. In addition, in view of the technological advantages possessed by the foreign party in joint-venture projects, an extremely important future question for China's policy-making institutions is how to make efficient use of the technological advantages of multinational corporations while simultaneously pursuing the health of 'independent development' and 'self-owned brands.'

In this chapter we have considered some problems related to vehicle transportation in China, evaluated the Chinese government's policy measures for alleviating these problems, and discussed some possible directions for future policies. As was true in the developed countries, China's policies will evolve toward a comprehensive structure involving laws, regulations, technological standards, fiscal subsidies, and tax incentives and penalties. However, in view of the great expectations of China's government and industrial community for 'independent development' and 'self-owned brands,' the formulation of technological policies related to vehicle transportation in China will be a complex game. Indeed, great expectations have been placed on the progress of China's domestic technology in general, including the cultivation of 'self-owned brands' and the establishment of technological standards in various other fields of industrial technology. Thus, even the most technologically advanced multinational corporations will not be allowed to persist in monopolistic exploitation of their technologies. Instead, an intelligent technological strategy for a multinational corporation would be to make efficient use of its own technological advantages while simultaneously taking an active role in the establishment of Chinese industrial technology standards, cooperating in the development of technologies and sharing technological benefits with Chinese companies.

Notes

1. Data source: http://www.stats.gov.cn/tjgb/ndtjgb/qgndtjgb/t20100225_402622945.htm (accessed May 11, 2011). The term 'civilian vehicles' excludes military vehicles, but includes all other publicly and privately owned vehicles. Three-wheeled and low-speed trucks, found primarily in rural areas and numbering about 13.31 million, are not included in the 62.88 million figure.
2. Including state-financed vehicles would give a higher rate (see the reference of Note 3). These state-financed vehicles have been widely used as private cars, in addition to their official uses – a source of significant controversy in Chinese society.
3. Wang Xiaoguang, Institute of Economics, National Development and Reform Commission; *China Automotive News*, July 22, 2007.
4. Industrial Economy Research Department of Development Research Center of the State Council, Society of Automotive Engineers of China, and Volkswagen Group China (2008) *Annual Report on Automotive Industry in China*, pp. 110, 131.
5. Vehicles in China are classified as light-, medium- or heavy-duty based on weight. Vehicles are then further categorized as M1 or M2 (for passenger cars) and N1 or N2 (for trucks) according to weight and number of passenger seats. 'Light-duty vehicles,' as specified in emissions regulations, refers to passenger cars (M1), buses (M2) and trucks (N1) with weight less than 3.5 tons. Passenger cars and trucks other than these are medium- or heavy-duty vehicles. In addition, Class M1 refers to passenger cars with weight of 1.0 ton and fewer than nine seats; Class M2 refers to passenger cars with weight of 1.0 to 5.0 tons and fewer than nine seats; and Class N1 refers to trucks with a weight of 1.0 to 3.5 tons.
6. By 2009, taxes and charges on vehicles in China included consumption taxes at production, purchase taxes at sale, vehicle and vessel taxes, compulsory traffic accident liability insurance and vehicle inspection costs during ownership, and fuel consumption taxes during operation.

7. *Beijing Daily*, July 28, 2008.
8. *National Business Daily*, July 9, 2007.
9. It was recently announced that future standards will regulate the average fuel efficiency of all vehicles sold by a given company, as in the US CAFE standards, instead of the efficiency of single vehicles (http://www.chinanews.com.cn/auto/2010/06-29/2369604.shtml) (accessed May 11, 2011).
10. http://cpsc.mep.gov.cn/gwgg/201002/W020100225545523639910.pdf (accessed May 11, 2011).
11. There are large numbers of trucks in rural areas, which are generally not included in Chinese automobile statistics. Most of these are three-wheel or four-wheel diesel vehicles. By the end of 2007, the population of such vehicles numbered about 17.4 million.
12. National High-Tech R&D Program proposed by several famous scientists in March 1986.
13. *Xin Hua Net* (http://www.xinhuanet.com/), September 19, 2005.
14. Zhu and Qu (2006).

References

Industrial Economy Research Department of Development Research Center of the State Council (DRC), Society of Automotive Engineers of China, and Volkswagen Group China (2008) *Annual Report on Automotive Industry in China*, Social Sciences Academic Press (China), April, 2008 (in Chinese).

Xi, J. and Chen, G. Q. (2006) 'Exergy Analysis of Energy Utilization in the Transportation Sector in China', *Energy Policy*, 34(14): 1709–19.

Jin, Y. (2005) 'Challenge of Fuel Consumption Faced by the Passenger Vehicle', *World Auto*, 1: 94–5 (in Chinese).

Kii, M. (2007) 'Nenpi kisei donyu seisaku no shouhisha beneki eno eikyo bunseki' (Analysis of Automotive Fuel Economy Standards Impact on Consumer Economic Benefits), *Jidosha Kenkyu* (*Japan Automobile Research Institute's Research Journal*) 29(2): 11–14 (in Japanese).

Kobos, P. H., Erickson, J. D., and Drennen, T. E. (2003) 'Scenario Analysis of Chinese Passenger Vehicle Growth', *Contemporary Economic Policy* 21(2): 200–17.

Li, X. M. (2005) 'Policy on Controlling the Emission of Pollutants by the Vehicle in China', available at http://www.auto.people.com.cn/GB/3545925.html (accessed November 15, 2010) (in Chinese).

Muto, S., Tokunaga, S., and Okiyama, M. (2006) 'Jidosha kanren kankyo seisaku no nisankatanso haisyutu sakugen koka keisoku; toppu rannar hoshiki to gureen zeisei o taisho to shite' (Evaluation of Regulatory Effects of CO_2 Emission by Implementing Environmental Policies Related on Automobiles: For Mileage Regulation by Top Runner Method and Green Taxes), *Chiikigaku Kenkyu* (*Studies in Regional Science*) 36(3): 683–96 (in Japanese).

Seko, T. (2007) 'Jidosha o torimaku haisyutu gasu oyobi nenpi no kisei doko'(Current Status and Future Trends of Emission and Economy Regulations), *Jidosha Kenkyu* (*Japan Automobile Research Institute's Research Journal*) 29(5): 3–8 (in Japanese).

Sun, L. (2007) 'Chugoku no jyoyosha shohizei kaitei no inpakuto ni kansuru CGE moderu shumireeshon bunseki' ('The Impacts of Consumption Tax

Reform for Passenger Car in China: Simulation Analysis by CGE Model'), *Kokusai kaihatsu foramu* (*Forum of International Development Studies*) 33: 89–98 (in Japanese).
Sun, L., Muto, S., Tokunaga, S., and Okiyama, M. (2006) 'Chugoku ni okeru jidosha kanren no kankyo, enerugii seisaku no teiryo bunseki: dogakuteki oyo ippan kiko (DCGE) moderu ni yoru hyoka' ('Numerical Analysis of Environmental and Energy Policies Related on Change "on" to "to"?Automobiles in China: Evaluation by Dynamic Computable General Equilibrium Model', *Chiikigaku Kenkyu* (*Studies in Regional Science*) 36(1): 113–31 (in Japanese).
Wang, C., Cai, W., Lu, X., and Chen, J. (2007) 'CO_2 Mitigation Scenarios in China's Road Transport Sector', *Energy Conversion and Management* 48(7): 2110–18.
Wu, W. (2005) 'Research and Establishment of the Standard on the Hybrid Electric Vehicle in China', *Communications Standardization*, 7: 8–11 (in Chinese).
Yokota, T. (2004) *Appendix: ITS Applications Around the World*, Washington, DC: World Bank, available at http://www.worldbank.org/transport/roads/its%20 docs/Appendix.pdf (accessed November 15, 2010).
Zhang, S. L. (2009) 'Introduction of the Technologies of New Energy Auto and Energy-saving in China', *Commercial Vehicle* 6: 26–9.
Zhang, S., Jiang, K., and Liu, D. (2007) 'Passenger Transport Modal Split Based on Budgets and Implication for Energy Consumption: Approach and Application in China', *Energy Policy* 35(9): 4434–43.
Zhu, H. and Qu, G. (2006) 'Architecture of Shanghai Transport Comprehensive Information System', 2006 Essays of the Second Chinese Intelligent Transportation Symposium, available at http://www.scctpi.gov.cn/thesis.asp?info_id=31&info_parentid=8&news_id=265 (accessed November 15, 2010) (in Chinese).

Appendix: Vehicle Categories and Specifications

Japan's Road Transport Vehicle Act classifies vehicles (passenger cars and trucks) into three broad categories: *ordinary-sized*, *small-sized*, and *mini-sized* vehicles. *Mini-sized vehicles* are vehicles whose length, width, and height do not exceed 3.40 m, 1.48 m and 2.00 m, respectively, and whose engine displacement does not exceed 0.660 liter. *Small-sized vehicles* have length, width and height less than 4.70 m, 1.70 m, and 2.00 m, respectively, and displacement less than 2.00 liters. Vehicles that do not fit into either of these categories are considered *ordinary-sized vehicles*. (Mini-sized vehicles are also known as *K-cars*.)

The Emission Control Standards define vehicle category for passenger cars as well as for trucks and buses. For passenger cars, the definition is the same as that in the Road Transport Vehicle Act. Trucks and buses are classified into three major categories: *light-duty*, *medium-duty*, and *heavy-duty* vehicles. *Light-duty vehicles* are defined as those whose gross vehicle weight is no more than 1,700 kg. *Medium-duty* vehicles are vehicles that do not fit in the *light-duty vehicles* category, but have gross vehicle weight no more than 3,500 kg. *Heavy-duty* vehicles are vehicles whose gross vehicle weight exceeds 3,500 kg. Trucks that have the same specification as *mini-sized* vehicles in the Road Transport Vehicle Act are classified as *mini-sized* vehicles.

In this book, for discussions and case studies of automobiles in Japan, we define three vehicle size categories, namely, *ordinary-sized*, *small-sized* and *mini-sized* vehicles, corresponding to the categories defined by the Road Transport Vehicle Law. We also define four categories of trucks, namely, *mini-sized*, *light-duty*, *medium-duty*, and *heavy-duty* trucks according to the Emission Control Standards.

For vehicle categories and specifications in China, we refer the reader to the notes of Chapter 8.

Index